ANIMAL
AS
MACHINE

ANIMAL

AS

MACHINE

∾

The Quest to Understand
How Animals
Work and Adapt

MICHEL ANCTIL

McGill-Queen's University Press

Montreal & Kingston • London • Chicago

ISBN 978-0-2280-1053-1 (cloth)
ISBN 978-0-2280-1221-4 (ePDF)
ISBN 978-0-2280-1222-1 (ePUB)

Legal deposit second quarter 2022
Bibliothèque nationale du Québec

Printed in Canada on acid-free paper that is 100% ancient forest free
(100% post-consumer recycled), processed chlorine free

We acknowledge the support of the Canada Council for the Arts.

Nous remercions le Conseil des arts du Canada de son soutien.

Library and Archives Canada Cataloguing in Publication

Title: Animal as machine : the quest to understand how animals work and
    adapt / Michel Anctil.
Names: Anctil, Michel, 1945- author.
Description: Includes bibliographical references and index.
Identifiers: Canadiana (print) 20210388560 | Canadiana (ebook) 20210388579
    | ISBN 9780228010531 (cloth) | ISBN 9780228012214 (ePDF) | ISBN
    9780228012221 (ePUB)
Subjects: LCSH: Physiology—History. | LCSH: Physiologists—History.
Classification: LCC QP21 .A53 2022 | DDC 571.109—dc23

This book was designed and typeset by studio oneonone in Minion 11/14

Contents

ঔ

Prologue

∾

The theme of functionality has haunted humankind for ages. As we extricated ourselves from the Stone Age, conjuring up objects to accomplish needed functions became part of the design process that surrounded us with tools and clever mechanical devices. The purpose of designing functional objects out of unformed matter in our environment, while it may have facilitated our survival as a species, served principally to alleviate the difficulties of living and ease the challenges of daily tasks; in short, to improve the quality of lived lives. Functional design became manifest in myriad shapes, in multiple domains of human experience. Ultimately, humans appropriated materials from their proximal environment to evolve a cultural extension of themselves.

The activities and shapes of the bodies of humans and other animals have been considered by many, past or present, to be testimonies to functional design. It is as if the functionality of hand-built objects emulates the functionality evolved within the living organism. When artisans create useful objects, they take into account not only their projected function but also their shape: how can it assist function? Preoccupations of this sort are embedded in the history of biology. Edward Stuart Russell (1887–1954), a British fisheries scientist, and philosopher of biology on the side, brilliantly expressed the ambivalence of the relationship between shape and functionality in his book aptly titled *Form and Function* (1916).

Russell asked: "Is function the mechanical result of form, or is form merely the manifestation of function or activity? What is the essence of life – organisation or activity?" Russell, who held holistic, anti-mechanistic views and was skeptical of Darwin's theory of evolution, sided with the like-minded

French comparative anatomist Georges Cuvier (1769–1832) in putting "function before structure." Russell emphasized that, in Cuvier's scheme, "structure and function are bound up together": "[E]very modification of a function entails therefore the modification of an organ. Hence from the shape of one organ you can infer the shape of the other organs if you have sufficiently extensive empirical knowledge of functions, and of the relation of structure to function in each kind of organ." Function determines the interdependence and adaptability of organs. If form and function are harmonized, they serve a common purpose; without that harmony, of what use is an organ?

For all the sense that Russell's embrace of Aristotelian teleology seems to make, his was an anatomist's view of a constructed animal world, a frozen world in which the dynamic underpinnings of animal actions were not questioned. He never asked the curiosity-laden question: how do organs actually accomplish their operations and by what specific mechanisms? In short, how do animals work? The following chapters will follow the bumpy and meandering historical trail along which concepts about animal functions were developed – the trail that led to the gradual emergence and subsequent thriving of the field of animal physiology.

In the formative years of human or medical physiology, animal physiology went through several stages of gestation. It could be said that, without the necessary "obstetrical" care or foresight on the part of natural historians, many of its potential offspring aborted or arrived stillborn. In this context, as the title of chapter 1 indicates, the study of animal functions remained an unacknowledged branch of natural history until the end of the eighteenth century. As the practice of basic physiology matured in the first half of the nineteenth century, the more attentive nurturing by devotees of animal physiology increased the discipline's chances of survival. However, these beginnings, as chapter 2 shows, were still uncertain. To ensure its emergence, animal physiology needed to be grounded in fertile environments in which schools of practitioners with shared concepts of animal functions could flourish. As outlined in chapter 3, a more united endeavour took root in the second half of the nineteenth century and the turn of the twentieth, a time dedicated to building sound foundations. On the basis of these collaborations, animal physiologists set out to deliberately fashion a discipline for themselves in the first half of the twentieth century. As shown in chapter 4, pivotal personalities, comprehensive books, and dedicated scientific journals served as magnets that drew together a cohesive body of knowledge and interest around which the discipline could take shape.

A measure of comparative physiology had already arisen in France by the first decades of the 1800s out of zoologists' desire to know whether "lower" animals function like humans and other mammals. It was not until the period from 1930s forward, however, that the implicit goal of these French zoophysiologists – to find common physiological principles among the diversity of animal forms – could be fulfilled. The processs took place differently in different countries, with the limelight shifting from England, France, and Germany (chapters 2–4) to countries like the United States (chapter 5), Belgium (chapter 6), and Canada (chapter 7), which infused new styles and a fresh dynamism into the field. The modern principles of comparative animal physiology – how functional unity emerges out of animal diversity; how individual animal species found innovative solutions to their specific problems; how animal functions adapt to their environment; and how animal functions evolved over time – were laid out in those years. Simultaneously, as chapters 6 and 7 recount, comparative biochemistry emerged and became integrated with the mission of animal physiology.

As the physiological/biochemical challenges met by animals in the wild became the focus of intrepid physiologists who led or participated in targeted expeditions to ecosystems around the world (as mentioned in chapter 7), it soon became evident that animals exposed to extreme environments accomplished extraordinary feats. Chapter 8 tells the stories of some of the men and women who brought these physiological heroics to light.

In the twentieth century, animal physiologists all too often concentrated on the study of functions they could assess with relative comfort: feeding, digestion, metabolism, breathing, circulation, renal function, and so on. For many animal physiologists, the nervous and endocrine systems were never on their radar, did not fit into their vision of what the field was about, or seemed too intractable or unworthy of the investment for their returns in explaining how animals work. One physiologist who sought to remedy this situation was the American C. Ladd Prosser (see chapter 5). In the fourth edition of his now-classic book *Comparative Animal Physiology* (Prosser

1991), he used the first volume to deal with the familiar sphere of the field, which he called *Environmental and Metabolic Animal Physiology*. But significantly, he devoted the second volume to *Neural and Integrative Animal Physiology*, by which he meant the ways in which animals sense their environment and how their nervous systems process sensory information and produce motor programs to execute appropriate actions mediated by muscle or other effector organs. He also included the ways in which hormones coordinate animal functions and behaviour. As the nervous system and the chemical system are clearly integral parts of animal physiology, chapters 9 and 10 of this book trace the history of discoveries and concepts that have recently become the bread and butter of comparative neurobiologists and endocrinologists.

Embedded in the story line of these chapters are four recurring motifs that have shaped the development of animal physiology: (1) technological innovations that have accelerated the pace of discovery by providing new equipment that allowed more precise measurements of physiological parameters, equipment often designed and built by physiologists themslves as the need arose; (2) the importance of the role played by marine laboratory facilities such as the network of marine stations in France, the Naples Zoological Station, the Plymouth Laboratory in the United Kingdom, and the Woods Hole Biological Laboratory in the United States, in providing a variety of marine animals and in-house research facilities to foster the comparative approach; (3) the diversity of approaches brought to bear on our understanding of how animals work: not only *bona fide* physiology, but also anatomy, biomechanics, biochemistry, ecology, animal behaviour, and evolutionary biology; (4) the value and importance of a collaborative *esprit de corps* in advancing comparative physiology as a discipline distinct from basic or medical physiology.

As several chapters of this book demonstrate, it was an uphill struggle to give the entire animal kingdom its due in physiological research. Zoophysiology, as it was initially called, was often overshadowed or met with disdain by medical physiology. This resistance began with medical scientists, for whom only laboratory or veterinary animals are worth considering. It continues with animal activists, whose vision of what is meant by "animal" is as narrow as that of the medical physiologists they challenge. As for the general public, the range of familiar animals often goes no further than traditional pets. Yet, this book shows that scientists studying how animals of all types work were as impressed by the usefulness of "exotic" animals for understanding the functioning of humans and other mammals as they were by the rewards of studying them for their own sake. Rewards also came in another form; no fewer than eighteen protagonists in this book, whether central or peripheral to the story, became Nobel recipients.

The compelling curiosity and passion that "zoophysiologists" poured into their work is laid out in this story. They stood in awe at the feats of animals as distinct as worms, bees, fish, and camels. Along the way their concepts of what their scientific discipline stood for was honed until it reached today's menu of environmental and molecular physiology. Many animal physiologists have excelled at bringing the strange functional adaptations of their animals of choice to life in their narratives. In his book *The Sense of Style* (2014), the cognitive psychologist and linguist Steven Pinker wrote: "I think about how language works so that I can best explain how language works." We can paraphrase him by asserting that animal physiologists have thought hard about how animals work so that they could best explain how animals work.

This book is not a comprehensive history of animal physiology. Several books on the history of physiology have treated it as basic or medical physiology; but the present book is to my knowledge the first attempt to provide a historical sweep of physiology through a zoological lens. Even then, my coverage is constrained by an abiding interest in how concepts about animal functions evolved. This means that several aspects of the field, its historical and more recent players, and the countries represented, have been neglected. I have had to make difficult choices and I stand by them. That said, I offer my apologies to all historians of biology and scientists in the field whose work or vision could not be accommodated in a book of this size.

∽

In preparing this book I was helped by my background knowledge and experience as a teacher of comparative physiology; but, more important, I relied on the amazing treasure of information for the scholar that the Internet now holds. Some of the documentary nuggets are buried in the vast, all-embracing web, and it took dogged detective work and a lifelong habit of research to uncover them. That done and the manuscript completed, experts pored over it, and I am grateful for their comments and criticisms, which greatly helped me to improve the content and form of this book. My editor at MQUP, Jacqueline Mason, guided and supported me with the unwavering competence and care that she has displayed in all our projects. Finally, my deeply felt gratitude goes to my copy editor, Jane McWhinney, who laboured over my three texts at MQUP, and numerous times rescued me from infelicitous turns of phrase, bumps in narrative flow, syntax errors, and typos. This and previous books owe much of their readability to her.

Chronology

∿

350–22 BCE Aristotle compares the body parts of different animal species and draws functional analogies. He also draws on biomechanical principles to describe animal limb movements.

162–217 CE Galen of Pergamon dissects animals killed by gladiators to discover how their internal organs are organized. As a precursor physiologist, he also experiments on animals to reveal how their organs are controlled.

1543 Andreas Vesalius, in *De humani corporis fabrica*, provides a detailed anatomy of the human body which lays the ground for the allocation of bodily functions.

1628 William Harvey uses the experimental method to shed light on the way the circulatory system works, and compares heart function in different animals.

1667 René Descartes articulates his representation of human and animal bodies as organic machines alongside the supernatural soul.

1668 Giovanni Borelli pioneers the study of biomechanics by showing how the limbs of humans, birds, fishes, and insects obey the laws of physics.

1747 Building on Descartes's idea, Julien Offray de La Mettrie shocks his readers by proposing that human and animal organs, even the brain, act like self-powered machines.

1755 Albrecht von Haller, who describes his physiology as *animated anatomy*, elaborates the notion that the functional responses of organs and tissues depend on their irritability.

1765–93 Lazzaro Spallanzani makes numerous experimental observations on a variety of animals, observations that propel him to a position as the foremost experimental biologist of his day.

1780–1800 Inspired by Spallanzani, Luigi Galvani and Alessandro Volta discover bioelectricity in frogs and other animals.

1800 Xavier Bichat makes the distinction between vegetative (involuntary) and animal (voluntary) functions.

1822–25 François Magendie makes the case for a dynamic experimental physiology free of functional deductions from anatomy. For him the activities of disparate organs are integrated into physiological systems.

1824–56 Marie-Jean-Pierre Flourens publishes the first comprehensive treatise on experimental neurophysiology. He also teaches the first course on comparative physiology, followed by *Principes de physiologie comparée* by Isidore Bourdon.

1826–40 Johannes Müller produces landmark studies on visual function in vertebrates and insects, and produces an influential textbook on human physiology.

1827 Henri Milne-Edwards introduces his theory of the division of physiological labour in more complex organisms.

1833 Henri Ducrotay de Blainville sees comparison of functions among animals as a search for common features indicative of the unity of life.

1838 Louis Antoine Dugès, in his treatise of comparative physiology, is the first to systematically cover the functions of invertebrates as well as vertebrates. Henri Milne-Edwards and his student Jean Louis Armand de Quatrefages follow in Dugès's tradition.

1842–52 Emil du Bois-Reymond launches the field of electrophysiology and compares muscle currents in a variety of animals.

1847 Carl Ludwig invents the kymograph to measure physiological activity in real time. His Leipzig laboratory attracts students from within and beyond Germany.

1865   Claude Bernard publishes his landmark opus, *Introduction à l'étude de la médecine expérimentale*. He introduces the notion of the constancy of the *milieu intérieur*.

1867   Hermann von Helmholtz publishes *Physiological Optics*, a milestone in the field of sensory physiology.

1870   Paul Bert, a student of Claude Bernard, produces a survey of respiratory functions in a variety of animals and launches the era of physiologists working at seaside marine stations.

1878   Etienne-Jules Marey publishes *La machine animale*, in which he creates ingenious methods of investigating animal locomotion.

1878   Marey's student Léon Fredericq discovers the octopus's haemoglobin, called haemocyanin, and goes on to build the Belgian School of comparative physiology.

1882–83   George Romanes, a friend of Charles Darwin, investigates animal intelligence and proposes a theory on its evolution.

1890–1918   Jacques Loeb studies animal tropisms, introduces an engineering approach to physiology, and, alongside Max Verworn, creates the field of general (cell) physiology.

1901   Georges Bohn introduces evolutionary physiology and coins the word "ethology," the study of animal behaviour.

1909   Keith Lucas pleads for the physiological study of a broad range of animals to gain evolutionary insights.

1913   Hermann Jordan publishes a book on the comparative physiology of invertebrates and becomes the first academic to take a chair of comparative physiology.

1913   Biochemist Lawrence Henderson introduces environmental physiology in his book *The Fitness of the Environment*.

1910–24   Hans Winterstein assembles prominent contributors to edit the highly influential eight-volume *Handbook of Comparative Physiology* in Germany.

1910–29   The Dane August Krogh makes important discoveries in comparative and medical physiology and becomes the first comparative physiologist to win a Nobel Prize. He is also known for the Krogh principle, according to which there is an animal model best suited for each physiological enquiry.

1913–54 Jean-Henri Fabre in insects and Karl von Frisch in fish discover animal communication by pheromones. Adolf Butenandt, Nobel Prize winner for the discovery of oestrogen, is the first to identify the chemical structure of a pheromone in the silkworm.

1921–47 Otto Loewi and Walter Cannon take the first steps toward the concept of neurotransmitters. Zénon Bacq compares chemical neurotransmission across animal groups.

1923 Lancelot Hogben and colleagues in Edinburgh create the *Journal of Experimental Biology*, destined to become an important outlet for comparative physiologists.

1924 Karl von Frisch and Alfred Kühn set up the *Journal of Comparative Physiology* in Germany.

1928–29 John Z. Young and Enrico Sereni make discoveries on the squid at the Naples Zoological Station which later lead to important advances in neurophysiology.

1928–37 Ernst and Berta Scharrer develop the concept of neurosecretion, stipulating that certain neurons produce and release hormones rather than neurotransmitters.

1929–62 Gottfried Koeller and Bertil Hanström posit the foundations of the comparative endocrinology of invertebrates. Aubrey Gorbman, an important figure in the development of the field, becomes the first editor of the journal *General and Comparative Endocrinology*.

1932 Joseph and Dorothy Needham pioneer comparative biochemistry in England, leading to *An Introduction to Comparative Biochemistry* by their student Ernest Baldwin in 1937.

1933–64 Cornelis Wiersma uses the crayfish model to pioneer the field of cellular neurobiology. He introduces the concept of the command neuron.

1937–81 Konrad Lorenz becomes the leader of the field of ethology, leading to his Nobel award.

1939–51 Clifford Ladd Prosser makes important contributions to invertebrate and fish physiology, and writes the first influential comparative physiology textbook in the United States.

1941–92   Donald Griffin unravels the sensory physiology behind echolocation in bats and publishes popular and controversial books on animal awareness and intelligence.

1944   Marcel Florkin, a product of the Belgian School, publishes *L'évolution biochimique*, in which comparative biochemistry is cast in evolutionary thinking.

1945–66   William Hoar, a student of the behaviour and endocrinology of salmon, becomes the leading comparative physiologist in Canada, his work culminating in his textbook *General and Comparative Physiology*.

1946–93   Theodore Bullock embarks on a vast research program of comparative neurophysiology that inspires new generations to select animal models aimed at understanding the role of specific neurons in locomotion, feeding, and other activities.

1947   Frederick Fry publishes the Canadian classic *Effects of the Environment on Animal Activity*, in which factors such as temperature are shown to determine survival and metabolic scope in fishes.

1947–58   Knut Schmidt-Nielsen and his wife, Bodil (daughter of August Krogh), study the physiological challenges of life in desert environments. He heralds the problem-solving approach of the engineer in his praised book *How Animals Work*.

1949–51   Arthur Hasler discovers the role of olfactory imprinting in the homing behaviour of salmon.

1955–72   Lawrence Irving and Per Scholander make pioneering contributions to the field of physiological adaptations to cold climates and diving, with Gerald Kooyman and Yves Le Maho following in their footsteps.

1958   George Bartholomew promotes the emerging field of ecological physiology.

1968–92   James Childress develops a unique research program on the physiology of deep-sea animals, culminating in his physiological inquiry of hydrothermal vent communities.

1972–2007 Bernd Heinrich discovers how bumblebees manage thermoregulation and their energy budget, an outstanding
contribution to behavioural/ecological physiology.

1973 Peter Hochachka and George Somero publish the paradigm-
changing *Strategies of Biochemical Adaptation*, in which they
underscore the fascinating adaptations of enzymes and other
proteins to challenging environments.

1976 Eric Kandel and his collaborators at New York University
Medical School use the sea hare *Aplysia* to launch a quest for
the neuronal mechanisms of learning and memory, later
awarded a Nobel Prize.

ANIMAL

AS

MACHINE

1

An Unspoken Branch
of Natural History
∾

Democritus says that we are foolish to imagine that we are superior to
animals since they have been our teachers in many arts: the spider in weaving,
the swallow in house-building, and the swan singing.
~ Gordon Lindsay Campbell (2014)

From the moment in prehistoric times when humans became conscious of
their separateness from other primates, their observations of animals have
evoked all manner of responses. What slivers of insight we may have gained
into the way Paleolithic humans viewed animals are derived from the great
documents of 30,000 years ago: cave paintings. As sociologist Linda Kalof
explains in her book *Looking at Animals in Human History* (2007), there are
abundant theories to account for these artistic representations of animals.
There seems, however, to be a consensus that such artistic renditions "were
expressing admiration for animals and most scholars agree that the prehistoric art work is closely linked to ritual and ceremony" (Kalof 2007). The
esteem in which these early animals were held extended to the qualities
displayed in their depicted activities: horses, bison, and even birds frozen in
mid-flight suggesting power and speed, for instance. This esteem implied
respect. But beyond these portrayals of the animal spectacle, Kalof argues,
the artists, especially in the Chauvet cave of Southern France, gave evidence
of certain knowledge of the external anatomy, behaviour, and social interactions of their animal models.

Admiration yielded in part to domination as humans became hunter-gatherers and eventually farmers. Neolithic farmers selected smaller, submissive, and hardy animals for breeding purposes. It is likely, for instance, according to Kalof, that about 12,000 years ago, the wolf became the first domesticated animal used for purposes other than breeding or food source; wolves not only assisted hunter-gatherers in hunting sorties but also served as "foragers of animal debris" and became objects of affection. Before the Graeco-Roman period, farmers started domesticating bulls to harness their muscle power for tasks beyond the scope of humans. In so doing, they faced ambivalence: they forced cattle into submission while at the same time admiring their strength to the point of devoting religious cults to them or honouring them as a source of emulation for the masculinity of tribal leaders. This respect for animal power would later translate into the practice of pitting the likes of lions against gladiators in Roman arenas.

What served as an entertainment centre for the Roman citizenry was also a source of occasional animal sacrifice for the nascent science of anatomy. Galen, the great Roman physician and anatomist (of whom more below), would go the distance to get his hands on any type of animal to satisfy his irrepressible drive for dissection (or even vivisection) as a way of overriding the ban or taboo on human dissection (Mattern 2013). As the physician attending gladiators in the amphitheatre of Pergamo, Galen apparently savoured his opportunities to dissect the exotic animals – imported from the far outposts of the empire – slaughtered by these gladiators (J. Donald Hughes in Kalof 2007). Galen's animal dissections often became bravura displays savoured by the numerous spectators as fond of gore as the audience for the gladiators. This was an inauspicious birth for the anatomical sciences, and it portended a disregard for the animal subject.

But how did the ancients generally approach the animal world if their ambition was to unravel the layout and inner workings of animal bodies? When ancient scholars turned their attention to animals as objects of study, in reality they were indulging in introspection. Their gaze was like a fishing line thrown out to the "creatures" in hopes that it would catch something telling about their own human make-up. They were interested in knowing the form and function of the body parts of animals only to satisfy their curiosity about the similarities or differences these parts presented in relation to the human body. The human body was the inescapable reference point. From time immemorial humans had a vague consciousness of the functioning of their bodies through the experience of life. As the French philosopher of science Georges Canguilhem (2008) explained:

> [K]nowledge of the functions of life has always been experimental – even when it was fanciful and anthropomorphic. For us, there exists a basic kinship between the notions of experiment [experience] and function. We learn our functions over the course of experiences and our functions then become formalized experiences. And experience is first and foremost the general function of every living being, that is, its debate with its milieu.

Despite this anthropocentric approach to the appreciation of nature, however, the intrinsic value of observing animals was not entirely lost on the ancients. Naturalists, in their all-encompassing interest in natural history, have always felt free to express their wonderment at the workings of animal bodies. Curiosity about animal functions manifested itself in early antiquity, either as a fascination with the peculiarities of animal activity or through the window of opportunity offered by comparing diseased with normal bodily functions. So, behind the question of how the human body functions, there often lurked the corollary question of what the diseased state can teach us about the role of the different organs of the body. And because trying to find answers by means of human experimentation presented ethical or technical challenges, learning how humans work soon became inextricably linked with learning how animals work.

For the curiosity-driven scholars of antiquity there was no clear boundary between natural history, physiology, and medicine in the sense we give these terms today. The word *physiologia* has a long history, having entered the vocabulary of the *savants* before the days of Galen of Pergamon (129–199), whom science historian Charles Singer considered to be "the greatest of the ancient physicians [after Hippocrates], and one of the greatest biologists of all time" (Singer 1957). As Vivian Nutton (2012) has explained: "One should be careful about assuming that when an ancient or a Renaissance doctor talked in terms of physiology he understood by it what we mean today. It is well known, certainly among classicists, that, for the most part, when the Greeks used the word *physiologia* and its cognates, they were referring not to a branch of medicine but to an investigation into nature as a whole."

The Greeks couched what little they learned from anatomical explorations of animals in a language that betrayed their vitalist outlook. For Empedocles (~ 480 BCE) body heat is maintained by the blood; and the heart, as the hub of the vascular system, distributes the *pneuma* – meaning at once soul and life – to the entire body through the blood vessels (Singer 1957). The Greeks believed that the proportion of fire and water in the body determines mood, temperament, sex, and intellect. Over a century later, Aristotle (384–322 BCE) expressed his vitalism differently, assuming that the heart is the seat of intelligence and the brain a coolant of the heart through a secretion he called *pituita* (Singer 1957). The pituitary gland derives its name from the fluid which was thought by Aristotle to pass "from the brain via the pituitary body and into the nasal passages" (Chester-Jones et al. 1987). In his work *De anima*, Aristotle separates bodies of the Earth along a simple dichotomy: bodies without *psyche* (inanimate) or with *psyche* (animated or living). He defines living bodies as having "the power of self-nourishment and of independent growth and decay" (Singer 1957), and he distinguished three types of *psyche*: vegetal (nutritive, reproductive), animal (sensitive), and rational (intellectual).

Aristotle's animal psyches may reflect many of what are now considered animal functions. In *De partibus animalium* [On the Parts of Animals], his idea of how animals work is permeated with teleological thinking. An animal's life has a purpose, he says, but to achieve this purpose it is necessary to put in place functions that serve that end. And even then, some functions are preconditions for the existence of others. Aristotle recognized, for example, that nutrition must come first to spur growth and sustain the development of other functions in the body, such as muscle activity and reproduction. Jason Tipton (2014), who delved deeply into Aristotle's biological thought, commented that the latter's consideration of the need to achieve a purpose "turns living bodies into tools." In this regard, Tipton analysed Aristotle's comparison of mouth parts in a variety of animals and showed how Aristotle viewed these parts as tools to an end – that of fulfilling nourishment needs by feeding methods suited to their kind. Sometimes very dissimilar animals made strange bedfellows, such as the cuttlefish and birds who, according to Aristotle, share a beak-like mouth.

It is perhaps worth pausing here to enlarge on this notion of biological tools. How far back in time must one travel to find observers or scholars musing on analogies with mechanical contraptions or domestic tools in order to explain how a body part operates? Or had humans always looked at biological materials as inspirations for crafting their useful objects? We cannot say, as the only documentary evidence we have, as far back as Graeco-Roman antiquity, is in print alone. It is only normal that observers should fall back on what is part of their everyday living to satisfy their curiosity about the functional capabilities of animals around them. That Aristotle indulged in this kind of exercise is made clear in this passage from *De motu animalium* [On the Movement of Animals], in which the motion of animal limbs is likened to levers by analogy with marionettes:

> The movement of animals resembles that of marionettes which move as the result of a small movement, when the strings are released and strike one another or a toy-carriage which the child that is riding upon it himself sets in motion in a straight direction, and which afterwards moves in a circle because its wheels are unequal, for the smaller wheel acts as a centre, as happens also in the cylinders. Animals have similar parts in their organs, namely, the growth of their sinews and bones, the latter corresponding to the pegs in the marionettes and the irons, while the sinews correspond to the strings, the setting free and loosening of which causes the movement. (Cited in Becchi 2009)

Similarly, Galen, in Book One of *On the Natural Faculties* (Brock 1916), by elucidating how ducts (ureters) carry urine from the kidney to the bladder and from the bladder to the outside through another duct, made an explicit analogy with water reservoirs in which flow and water level can be controlled. Likening lungs to bellows is another example. However, Galen never went so far as to reduce body parts to simple machines. He explicitly acknowledged that bones, muscles, blood vessels, and other components, unlike man-made machines or tools, are capable of growth and plasticity as they go on functioning, thereby making their *modus operandi* a lot more complex.

Aristotle examined a variety of animals and organized them on a scale of organic complexity that suggests his use of a crude comparative method. This approach led him to discover a peculiar feature of sharks, as Singer (1957) explained:

> Perhaps his most extraordinary anatomical feat is his account of the placental development of the dogfish *Mustelus laevis* ... Aristotle ... paid special attention to the habits and structure and especially the breeding of fish. He knew that they were mainly oviparous, but occasionally viviparous, and he knew also of one instance among the Elasmobranch fishes (which he called Selachia), in which the development bore an analogy to that of placental mammals. This fact remained almost unnoticed until the nineteenth century, and it was its rediscovery that drew the attention of naturalists to the great value and interest of the Aristotelian biological masterpieces.

Beyond the intrinsic value of his comparative standpoint, which allowed the discovery of a shark's placenta that preceded the existence of placental mammals by hundreds of thousands of years, it is useful to ponder Aristotle's method of observation. He did not so much follow the dynamics of an animal's functional organs as observe anatomy and make deductions on function by analogy. The reference point was anatomy by dissection, and function was never ascertained by experiment. It would never have occurred to Aristotle and his peers to make a conscious distinction between anatomy and physiology; to study anatomy was at the same time to study function. It was far from a foolproof method. As Singer noted: Aristotle "made no proper distinction between arteries and veins, and he believed (quite incorrectly) that arteries contained air as well as blood. He failed, too, to trace any adequate relations between the sense organs, the nerves and the brain."

The generation following Aristotle produced Erasistratus (~ 290 BCE), whom Singer considered "the Father of Physiology" – a hollow claim, contradicted by his own appreciation of the Greek's work. The constructs of function that Erasistratus produced were as speculative as those of his predecessors, and he certainly did not elevate physiology to the status of a discipline. Singer also claimed that Erasistratus was a rationalist, but the Greek anatomist's views of the heart, blood vessels, and nerves were rooted in the same vitalism of *pneuma* or *psyche* invoked by his predecessors. He did, however, stumble upon prescient insights, such as his view of nature as "a great artist acting as an external power shaping the ends to which a body acts" (Singer 1957), hinting perhaps at the role of the environment in shaping function. He also used the Aristotelian comparative method to observe that the human cerebrum possesses convolutions that other mammals lack, inferring that this observation explains humans' higher intellect.

With only a few exceptions, the vitalist outlook, relayed by Galen in the second century CE, remained the mainstay for the interpretation of animal functions until the seventeenth century. Even as observations or deductions of organ function came to depend on more reliable anatomical information, little advancement occurred until the appearance in 1543 of the great work of Andreas Vesalius, *De humani corporis fabrica* [On the Fabric of the Human Body].

Vesalius (1514–1564) was born in Brussels into the family of a pharmacist attached to the Imperial Court of Maximilian I (O'Malley 1964). His initial education at the University of Louvain was followed by medical studies at the University of Paris (1533–36) and then a return to Louvain (1536–37). At the age of only twenty-three, he was appointed professor of surgery at the University of Padua in Italy. Early in his tenure in Padua (1537–42), he completed a compendium of the "anatomical-physiological views" of his idol Galen, which served as a springboard to his exhaustive anatomical research on the human body published in Basel the year after he left Padua. Unprecedented at the time, Vesalius's treatise was richly and superbly illustrated and relied entirely on his own dissections rather than deferring to previous authority. After the publication of *Fabrica*, Vesalius was appointed Imperial Physician to Charles V and he never returned to academic life. He died in Greece on his way back to Europe after a pilgrimage to Jerusalem.

Vesalius's detailed and largely accurate descriptions of the human body went a long way toward laying the ground for the allocations of bodily functions that began to roll out in the seventeenth century. Vesalius himself was no "physiologist," and Foster (1901) remarked that *Fabrica* "is in the main a book of anatomy," adding that "the physiology is incidental, occasional, and indeed halting." But what physiology? Numerous scholars have addressed the historical development of physiology in the sixteenth and seventeenth centuries, but their efforts were muddled by a poor grasp of the

science as practised. Historian of science Andrew Cunningham has disen-
tangled the mesh of confusion by showing that the physiology of the 1500s,
1600s, and 1700s, which he called "old physiology" or anatomical physiology,
had nothing to do with the experimental physiology that emerged at the
cusp of the nineteenth century (Cunningham 2002). The old physiology, in
fact, rather than being a scientific discipline as Cunningham seems to believe, looked more like a theoretical construct in the mind of observers of
anatomical features.

Jean Fernel (1497–1558), the French physician to Henri II, provided a prime
example of this so-called physiology. Indeed, Sir Charles Sherrington, the
famous neurophysiologist, and others even more recently (Welch 2008,
Tubbs 2015), tried to herald Fernel as the first experimental physiologist
(Sherrington 1946). Cunningham (2002) convincingly debunked this fantasy.
Fernel himself referred to his status as that of a "médecin philosophe" (Figard
1903), and his theorizing was considered to be no more than a branch of natural philosophy. To claim that Fernel founded the discipline of physiology
because the word *physiologia* appeared in one of his tomes seems preposter-
ous. Until close to the nineteenth century, anatomy was *the* discipline, and
any interest in studying function stemmed uniquely from the observation
of structures. That said, a few investigators did manage to include genuine
experiments in their work, and the two most eminent examples from the
seventeenth century, William Harvey (1578–1657) in England, and Giovanni
Borelli (1608–1679) in Italy, are cases in point.

William Harvey, the son of a businessman and local politician, was born
in Folkestone, a coastal village by the English Channel (Shackelford 2003;
Wright 2012). He studied clinical medicine at the University of Padua, however, where the likes of Vesalius, Gabriele Fallopio (of Fallopian tube fame),
and Hieronymus Fabricius (discoverer of the venous valves) had taught or
were teaching. Upon Harvey's return to England in 1602, his medical practice
flourished and his functions at the Royal College of Physicians gave him visibility among the rich and famous of London. He also became a lecturer of
surgery, and "he supplemented his human dissections with comparative
anatomical demonstrations of lower animals. Harvey's use of comparative
animal research was part of his program" (Shackelford 2003). It was during
these years that he conducted his research on the circulatory system leading

to his publication of *Exercitatio anatomica de motu cordis et sanguinis in animalibus* (An Anatomical Exercise on the Motion of the Heart and Blood in Living Beings) in 1628. Shackelford situates Harvey's contribution in the context of the scientific revolution:

> This revolution was part of a broader transformation of scientific thinking and practice that completely rearranged the intellectual world of Europe. The dramatic change, which historians have called the Scientific Revolution, did not occur in a vacuum, but was shaped by the historical context in which Harvey and his contemporaries worked. The history of William Harvey's medical speculations and of his experimental approach to scientific argumentation reveals the beginnings of this revolution in medicine and biology.

Many scholars shared Shackelford's view of Harvey's work as a pivot of the scientific revolution but, according to Wright (2013), a closer examination of Harvey's texts would indicate that his approach was inspired as much by the old Aristotelian teleological way of thinking as by his own brand of the nascent scientific method. To make his point, Wright claims that Harvey "sedulously followed Aristotelian logic in concluding that, 'since nature does nothing in vain', and since 'so provident a cause as nature' could not have 'plac'd so many valves without design', the vessels and the valves had to have a purpose; ultimately, this purpose could only be the circulation of the blood."

Harvey's milestone was to have been the first to break away from Galen's teaching that blood was transferred between the two heart ventricles through leaky membranes and that venous blood flow was entirely independent of arterial blood flow. Harvey used anatomy as a guide to intuit how the blood circulates among the heart chambers and how it moves from the arterial to the venous systems. He remained puzzled, however, as to how veins are filled of blood by arteries, and the answer – through the capillary beds – only came more than thirty years later thanks to microscopic observations on frog lungs by Marcello Malpighi (West 2013a, Loriaux 2016). But Harvey's special genius consisted in designing experiments to test his anatomy-based hypotheses, which led him to conclude that blood is conserved and recycled in a circulatory loop from the heart and back. Stanley Schultz, a renowned physiologist and past president of the American Physiological Association, aptly summarized Harvey's experimental contribution involving measurements:

> First, he measured the total amount of blood that could be drained from sheep, pigs, and some other sub-primate mammals. He then measured the volume of the left ventricles of these animals and calculated that, if the left ventricle were to empty with each beat, in one hour the total volume of blood pumped would be much greater that in the ingesta or even that contained in the entire animal. Indeed, this would be true even if one-tenth of the blood contained by the ventricle were ejected per beat. Therefore, he concluded, "it is a matter of necessity that the blood performs a circuit, that it returns to whence it set out."
>
> He then demonstrated, publicly, that when a live snake is "laid open," compression of the vein entering the heart leads to a small heart that is devoid of blood upon opening it. (Schultz 2002)

From the perspective of today's animal physiologist, this passage shows that Harvey willingly embraced experimentation on animal species other than humans. It is implicit in his way of doing research that he pioneered the selection of animal models best suited to answer specific questions, a practice that was to become common to general physiology as well as one of the aims of comparative physiology in the twentieth century (see chapter 4). But more important, Harvey engaged also in a comparative analysis of heart activity, from lower (vegetative) animals in which no circulatory system is detected to insects and shrimps in which "there is a pulsating place like a vesicle or auricle without a heart [that] may be seen beating and contracting, slowly indeed, and only in the summer or warmer seasons" (Harvey 1628). These invertebrates are contrasted with non-mammalian vertebrates: "In larger, warmer, red-blooded animals there is need for something with greater power to distribute nourishment. So, to fishes, serpents, lizards, turtles, frogs and such like, a heart is granted with both an auricle and ventricle."

Much has been made of Harvey's experimental method, but he also spent time simply watching the live process of blood circulation – haemodynamics. This activity can be called descriptive physiology as opposed to experimental physiology. As a historian but not a practising scientist, Cunningham (2002) confused Harvey's descriptive physiology for an anatomical exercise. Anatomists can deduce organic functions from morphologies, but physiologists can learn about function from observing the activity of animal parts without necessarily resorting to experimentation, although the latter is always welcome, especially when a hypothesis needs to be tested. Harvey used the word "anatomy" in the title of his work because he saw himself as a professional anatomist, even though the work itself was largely – and perhaps unconsciously – physiological in the modern sense. If it were a work of anatomy, one would expect anatomical illustrations to support anatomical arguments, but the monograph contains only one figure – showing the use of tourniquets on the human arm to observe the effect on the blood circulation in superficial veins and arteries.

What Harvey stood for was an openness of mind that from antiquity to his time had been in short supply. He was conscious of his intellectual process and explained it eloquently in his 1628 opus:

> True philosophers, who are only eager for truth and knowledge, never regard themselves as already so thoroughly informed [that they do not] welcome information from whomsoever and from wheresoever it may come; nor are they so narrow-minded as to imagine any of the arts or sciences transmitted to us by the ancients, in such a state of forwardness or completeness that nothing is left for the ingenuity or industry of others. On the contrary, very many maintain that all we know is still infinitely less than all that remains unknown. [Nor] do philosophers pin their faith to others' precepts in such [ways] as they lose their liberty, and cease to give credence to the conclusions of their proper senses. Neither do they swear such fealty to their mistress Antiquity that they openly, and in sight of all, deny and desert their friend, Truth. (Quoted in Schultz 2002)

But Harvey's seemingly unassailable arguments left his colleagues and future biologists unmoved. Shackelford (2003) offered reasons: "[S]ome were unconvinced that experimental evidence was relevant to physiological research. Others doubted Harvey's ideas because they lacked proper grounding in metaphysics, the traditional philosophical principles that were used to explain the world." However, there were a few believers who based their own research on the new method and knowledge generated by Harvey. One of them was Giovanni Alfonso Borelli.

Born into the family of a humble Neapolitan soldier, Borelli (1608–1679) studied mathematics in Rome, and his great gifts were soon noticed among Italian academics, leading to his procurement of the chair of mathematics at the University of Messina around 1640 (Foster 1901). As his fame as a mathematician and physicist grew, he was invited in 1656 by a member of the Medici family, Ferdinand Duke of Tuscany, to fill the chair of mathematics at the University of Pisa, which Galileo had filled a few years earlier. Spurred by his ducal sponsor, Borelli embraced the experimental approach in his work during his Pisa years (Boorstin 1983). This period led to the writing of his ambitious treatise *De motu animalium* which was largely completed by 1668 when he left Pisa. The work was published only after his death, however, in 1680.

On the title page of his treatise, Borelli called the work a "physico-mechanical dissertation," thereby implying that principles of physics and mechanics had been brought to bear on the study of motricity in animals. In the introduction he made clear as well that principles of geometry were also applied to the shape and configuration in space of the components: muscles, tendons, ligaments, bones, and cartilage. Borelli is considered to be the founder of "iatrophysics," the application of physics to medicine that is today called biophysics (Boorstin 1983), but a reading of his treatise suggests rather that he pioneered the modern discipline of biomechanics. Boorstin summed up fairly well the reach of Borelli's innovative approach:

In this work Borelli showed that movements of the human body were like those of all other physical bodies. When a man's arm lifted a weight, the work was accomplished according to Archimedes' familiar principles: the bone was the lever, moved at its shorter segment by the pulling force of the muscle. Movements of the limbs in lifting, walking, running, jumping, and skating also followed the laws of physics. Borelli showed how the very same laws governed the wings of birds, the fins of fishes, and the legs of insects. Having explained the body's "external" motions in his first volume, he proceeded in his second volume to apply these same physical laws to the movements of muscles and the heart, the circulation of the blood, and the process of respiration.

Again, as Harvey before him, Borelli relied on anatomy for the layout of the components of the motor system, but the rest was not anatomy; in Borelli's case it was no less than an engineering approach that befitted his training. As such he made estimates of the forces generated by muscle activity and the direction of these forces in space, leading to specific limb movements. The end-result was that he followed the dynamics of living phenomena. And like Harvey he compared different animals so that he could gain insights into various modes of animal locomotion. He tried to answer such perplexing questions as "Why the birds rest and sleep standing on and grasping tree branches, without falling" or "How quadrupeds attempt to use their two forelegs like hands to grasp objects" (Borelli 1680).

On a more fundamental level, Borelli attempted to understand the mechanism of muscle contraction. According to Foster (1901), "he did not resolve how the muscle fibres produce contractions, even though the nascent method of microscopy allowed him to see the fiber arrangement inside muscles which is responsible for the contraction." However, he was the first to clearly expose the role of the nervous system – from the brain downstream – in activating muscle contraction, although he could only use the "fluid" or "juice" metaphor to describe signal transmission from nerve to muscle. As for muscle contraction itself, he saw its cause in a swelling and hardening of the muscle fibres.

Through his muscle research, Borelli gained a better understanding than his contemporaries of the breathing mechanism. He showed that: "The movement of inspiration is carried out by the intercostal muscles and the diaphragm acting simultaneously … During quiet and normal expiration air is not ejected by the force of some muscles but as a result of the quietness and absence of action of the intercostal muscles, of the relaxation of the diaphragm and of the opening of the epiglottis" (Borelli 1680). He also realized that a fair volume of air remained unevacuated from the lungs during expiration. He cited the works of the English physicist and chemist Robert Boyle (1627–1691) and the Italian mathematician and physicist Evangelista Torricelli (1608–1647) to press his point that breathing air was more critical to life than the heart and circulation: "The experiment which must obviously prove this assertion is the temporary removal of air in the pump of Boyle or, even better, in the vacuum of Torricelli achieved by using quicksilver [mercury]. All animals shut up in such a machine fall moribund immediately but, if air is carefully returned, the same animals come back to life."

Borelli was referring to Torricelli's 1644 invention of the barometer, a column in which mercury separated a vacuum space from atmospheric air (West 2013b). The higher the ambient air pressure, the greater the amount of mercury that is pushed into the vacuum space. But the decisive experiments were reported by Boyle in a monograph published in 1660 (West 2015) in which, curiously, the phrase *New experiments physico-mechanicall* figured in the title. It makes one wonder whether Borelli had borrowed it for the subtitle of his own work. The vacuum pump designed by Boyle with the help of his colleague Robert Hooke was an engineering feat at the time. In his "Experiment number 40," Boyle used insects (flies and bees) and small birds (larks) that could be accommodated in the small bell jar from which the air was evacuated. The insects dropped inanimate as soon as the air was largely gone, whereas the birds went through convulsions before collapsing. Nine years later, another Englishman, the physician Richard Lower, enlisted Hooke's artificial respirator to demonstrate that "air" from the lungs is transferred to arterial blood. Thomas Baskett recounts that in 1667, by "using a pair of bellows attached to a pipe in the trachea of a dog [Hooke] showed that as long as the lungs were inflated the motion of the heart was normal" (Baskett 2004). In 1669 Lower reported in his *Tractatus de corde* [A Treatise on the Heart] that, when air was pushed into the lungs of an open-chested animal by Hooke's respirator, the blood in the pulmonary vein, which goes to the heart, changed from a dark, venous-like colour to a "florid" colour (Lower 1728, Tubbs and coll. 2008). The experiment was the opening salvo in the process leading to the modern concept of blood oxygenation in the lungs.

∼

All these impressive achievements in the second half of the seventeenth century, in the wake of Harvey's founding contribution, would suggest that *bona fide* experimental physiology was being practised, if not as a general discipline, at least by occasional luminaries. As we turn to the Age of Enlightenment, it becomes apparent that physiology had made little significant progress over the previous century in the understanding of animal functions in any but the theoretical sphere. The *siècle des lumières* represented a dramatic shift of philosophical thinking that spilled over into the science domain in a major way. Unsurprisingly, the French led the way in the natural sciences also. A prominent representative of the new materialist trend that touched on physiology was Julien Offray de La Mettrie (1709–1751).

La Mettrie was born in Saint-Malo, Brittany, to a wealthy textile merchant (Vartanian 1960). After a brief flirtation with a vocation in the Church, he studied philosophy and natural science. Between 1728 and 1733 he was a student at the Medical School of Paris. For a year La Mettrie trained in Leiden with the renowned Dutch natural scientist and physician Herman Boerhaave (1668–1738). Boerhaave was famous for pioneering medical chemistry and clinical teaching, but his science, according to historian and biographer Rina Knoeff (2002), was heavily coloured by his staunch Calvinism. La Mettrie took the "clinical perspicacity and experimental leanings" he learned from Boerhaave back to Saint-Malo, where he established his medical practice, and he translated his master's work for dissemination in France (Vartanian 1960). In 1743 La Mettrie became a medical officer in the army and he experienced several battles of the War of the Austrian Succession. From 1745 to 1746 he served as medical inspector of army hospitals. But his satirical and vitriolic exposure of "the self-seeking ineptitude, which in the medical world of the eighteenth century, seemed to typify the French practitioners in particular," forced him into exile in Holland (Vartanian 1960). It was during his exile in Leiden, in 1747, that he wrote his foundational book, *L'homme machine* (La Mettrie 1865).

La Mettrie gave the most mechanistic answer the *siècle des lumières* could possibly offer for the perplexing question: how do humans and other animals work? *L'homme machine*, right from its explicit title, shocked its readers not only with its description of the human body as a physical machine, but also because La Mettrie did not mean it as a metaphorical shorthand. The French physician-philosopher was inordinately forceful in his argument that even mental processes are reducible to machinery. In his scheme, no God is necessary, whether or not through the intermediary of a soul. As Adam Vartanian (1960) explained, La Mettrie's "vehement tone and blunt style were

Portrait of Julien de la Mettrie. Engraving by Petrus Antipief.
National Library of Medicine Image B06066.

Portrait of Albrecht von Haller by Johann Rudolf Huber in 1736.
Public Library of Bern, Negative Number 2453.

largely due to the fact that *L'homme machine* was not only an exposition of materialist science but a powerful piece of philosophic propaganda as well – for these two aims were not easily separable in the Enlightenment."

The perception of living bodies as machines did not start with La Mettrie. Although several scholars had alluded to the concept in the past, only René Descartes (1596–1650) articulated it in a comprehensive system of thought. The author of *Discours sur la méthode* differed from La Mettrie in believing that an intangible soul followed rules other than those of the animal machine. Auguste Georges-Berthier (1888–1914), the brilliant specialist in Descartes's physiology who died on the battlefield in the first weeks of World War I, interpreted the Cartesian mindset as follows: "The soul is completely alien to organic phenomena; hence the expression of the Cartesian mechanism: the living body, as everything else in the universe, is a machine wherein all is done by figure and movement" (Georges-Berthier 1914). Descartes's idea of the organic machine had only come to light in a posthumous treatise titled *De l'homme et de la formation du foetus* (1677). In it Descartes reviewed the various tasks undertaken by the animal machine: food digestion, nutrition, growth, heartbeat, respiration, waking and sleep, reception to light, sound, smell, taste, and heat. Oddly, he viewed the pineal gland as the brain centre where these many sensory inputs converge to organize thought and memory. In the conclusion of his treatise, Descartes emphasized that these functions are accomplished by machinery no different from mechanisms designed by humans:

> Thus, I say, when you reflect on how these functions follow completely naturally in this machine solely from the disposition of the organs, no more nor less than those of a clock or other automaton from its counterweights and wheels, then it is not necessary to conceive on this account any other vegetative soul, nor sensitive one, nor any other principle of motion and life, than its blood and animal spirits, agitated by the heat of the continually burning fire in the heart, and which is of the same nature as those fires found in inanimate bodies.

La Mettrie largely agreed with Descartes's mechanistic credo, but his exploration of the mechanical animal went further than that of his compatriot of a century earlier. First, he took the soul out of the equation, substituting "the Method of those who would follow the path I open them, to interpret supernatural things, incomprehensible as such, by the lights one has received from Nature" (La Mettrie 1748). Second, La Mettrie examined with greater alacrity the cogs of the living machine. He argued that, contrary to man-made machines, which require an external force for activation, animal machines are self-powered. He attributed this empowerment to a machine's ability to rewind itself, arguing that this rewinding, or oscillation, is accomplished by the irritability of tissues. The Latin word *irritabilitas* had originally been co-opted by the English physician Francis Glisson (1599–1677) to define muscle contractility. But La Mettrie expanded its meaning to include excitability in other tissues as well. Nerves are irritable but so are the tissues targetted by the nerves. Nerves can excite a tissue or organ, but take out the nerves and a tissue such as muscle can excite itself: "[E]ach minute fibre, or component of organised bodies, moves by an intrinsic principle the action of which does not depend on nerves ... Such is this motor principle that entire Bodies or parts cut in pieces produce movements, not deregulated as believed, but very regular." The seat of this intrinsic power is found "in the substance proper of the component parts – without recourse to veins, arteries, nerves – in brief, in the organization of the entire body" (La Mettrie 1748).

The core of La Mettrie's originality lay in his concept that irritability is central to life processes and his notion that mental processes have a biological foundation. These ideas naturally ran counter to the religious beliefs held by his contemporary scientists. "So lively was the hostility aroused by the appearance of *l'Homme machine*," remarked Vartanian (1960), "that it ranks as perhaps the most heartily condemned work in an age that saw the keenest competition for such honors. As was to be expected, these attacks served merely to compound and prolong the book's *succès de scandale*."

Scandal it provoked indeed, and La Mettrie was forced to leave his Dutch haven and find refuge in Berlin, where Frederick the Great felt sympathetic to the Frenchman's unorthodox ideas. When La Mettrie died in 1751 from an apparent complication of a digestive disorder, it was the Prussian king who wrote the eulogy (Frederick II of Prussia 1752). Frederick, who had genuinely enjoyed the Frenchman's company, closed his eulogy with this assessment:

M. La Mettrie was born with an inexhaustible fund of natural gaiety;
he had a quick mind and an imagination so fertile that it grew flowers

in the arid ground of medicine. He possessed a natural talent for speech
and philosophy, but an even more precious gift from Nature was a pure
soul and a generous heart. All those whom the pious insults of the-
ologians leave cold mourn in M. La Mettrie an honest man and a medical scholar.

La Mettrie dedicated *l'Homme machine* to Albrecht von Haller (1709–
1777), the renowned Swiss physician who is often considered to have laid the
foundations of modern physiology but whom Cunningham (2002) associates with old physiology – or "animated anatomy," as Haller himself defined
his physiology (Haller 1751). Like La Mettrie, Haller studied under Boerhaave
in Leiden but retained from his mentor the Calvinist faith that separated him
from the Frenchman. In fact, Haller was outraged by La Mettrie's "impious"
physiological system (Haller 1755). His attitude reflected the cultural divide
of northern Europe at that time. "In building on the grounds of religion,"
explained Hallerian scholar Otto Sonntag (1974), "[Haller] mirrors the prevailing German and Swiss outlooks of the Enlightenment, an age in which
many Frenchmen paid the religious end of science lip service at most."
Haller's greatest original contribution was to have introduced the concepts
of irritability and sensitivity in the study of different organs and tissues
(Haller 1755). Experimenting on dogs, cats, goats, and frogs, he extended the
implication of his findings to humans:

> I call irritable any part of the human body which becomes shorter when
> some foreign body touches it a bit strongly. Assuming that the tactile
> force remains constant, the irritability of the fibre is greater if it shortens
> more. The fibre that shortens a great deal in response to a light contact
> is highly irritable; that which produces only a small shortening in re-
> sponse to a violent contact is slightly irritable.
>   I call sensitive any fibre in man which, when touched, transmits to
> the soul the impression of this contact: in animals, in which the pres-
> ence of a soul is uncertain, we shall call sensitive any fibre the irritation
> of which produces in them obvious signs of pain or discomfort. In con-
> trast, I call insensitive that which, suffering burns, cuts, piercing or
> worse, induces no sign of pain, no convulsion, no change in the state
> of the body. (My translation)

This passage clearly shows that Haller's "irritability" applied only to muscle contraction, in sharp contrast to La Mettrie's "irritability," which was a property of all tissues and went beyond contractility. In addition, it transpires that Haller called "muscles" those fibres that articulate to the skeleton such as limb muscles; but other organs which shortened – heart, iris, intestine, and so forth – were not explicitly identified by him as muscles, although they contract by similar mechanisms. Because Haller could measure changes in fibre length as a result of contractions, he was able to obtain experimental data in the modern sense of physiological inquiry, as Dominique Boury (2008) explains. His "sensitive fibres," on the other hand, were not as sharply defined as the irritable ones, and certainly not measurable. The theologically phrased "impression to the soul" seems to convey an idea of perception based on sensory input to the brain, but whether Haller meant it that way is not clear. What is clear is that for him, to call nerves sensitive meant that they have feeling. Muscles are sensitive as well as irritable owing to the presence of nerves in them, Haller claimed; but muscle irritability, he said, does not need nerves for its activity to unfold.

For Haller, muscles are sensitive only if nerves connect them to the brain/ soul. Nerves are unnecessary for muscle contraction; oddly, when Haller "irritates" nerves to a muscle, he is more interested in knowing whether the nerves contract or not – they don't – than in testing whether exciting the nerves causes muscle contraction. He failed to envision that, even though muscles can contract as a result of direct irritation, in their normal living condition – with few exceptions such as the heart – they need motor nerve input to contract. Marco Piccolino and Marco Bresadola, in their engrossing book *Shocking Frogs* (2013), relate that a "neuro-electric view of animal motion" existed in Haller's time but that Haller himself, in his *Elements of Physiology of the Human Body* (1762), rejected it on the grounds that the experimental data were inconsistent with such a theory. It fell to Luigi Galvani (1737–1798) to demonstrate that electric phenomena intrinsic to nerves and muscles are involved in motor activity.

Piccolino and Bresadola, in their narrative of the birth of bioelectricity (2013), give a comprehensive account of the kind of physiology practised in the late eighteenth century. It was closer to the imagery of the animal machine than the Hallerian model and, however haltingly, it presaged nineteenth-century physiology in the development of hypothesis- or criticism-driven experimentation. It also anticipated the importance of developing instruments and designing experimental protocols best suited to specifically address scientific questions. The spirit of the French Revolution was often said to account for this new approach to science (Schiller 1968). It was not in France, however, that the new scientific approach was embodied, but in Italy, in the intriguing and larger-than-life personality of Lazzaro Spallanzani (1729–1799).

Born in Scandiano in northeast Italy to a lawyer father, Spallanzani was schooled at home and by Jesuits before attending the University of Bologna (Rostand 1951). In Bologna he developed a lifelong attraction for the natural sciences. During his academic posts in Reggio, Modena, and particularly in Pavia, his eclectic curiosity drove him to investigate a host of biological and physiological problems: blood circulation, digestion, respiration, reproduction, regeneration, and the origin of germ organisms. In a letter to Spallanzani dated 29 November 1780, Swiss biologist Charles Bonnet conceded that his Italian colleague was making more discoveries "than entire Academies in a half century." Spallanzani, who (maybe rightly) thought highly of himself and his works (Castellani 1991), matured into what Jean Rostand (1978) regarded as the foremost pioneer of experimental biology:

> Spallanzani is not a theoretician, but a pure researcher. Instead of reasoning, philosophizing, or straining to imagine what is, as so many have done before him, he questions the facts directly; he possesses, as Pasteur will say later of him, "the experimental reflex"; and, on this score alone, one must regard him as one of the authentic founders of modern biology. (My translation)

Spallanzani's intellectual influence is noticeable in experiments by his fellow Italians Galvani, who openly admired him, and Alessandro Volta (1745–1827). Both Galvani and Volta contributed to the birth of electrophysiology, even as a historically important controversy erupted between them, and their research approach contained the seeds of what would become the subfield of comparative physiology.

Galvani studied medicine and philosophy at the University of Bologna, his birthplace. He remained at the university as an anatomist from 1762 to 1775, when he was appointed professor of anatomy and surgery. He was also affiliated with the local Academy of Sciences, which sponsored his research. "Galvani's research," Piccolino and Bredasola (2013) remarked, "flourished within a milieu characterized by the great interdisciplinarity of the scientific approach, the Institute of Sciences of Bologna, which was one of the first modern experimental institutions in Europe." Although Galvani benefitted from scholarly exchanges with colleagues at the academy and reported on his research results within its walls, his research on bioelectricity was actually conducted in a room in his home specially arranged for this purpose.

According to Piccolino and Bresadola, Galvani's bioelectricity research began in 1780, spurred by recent reports on electric fish (torpedo and electric eel) and by the discussion in his lectures of the "neuro-electric theory of vital functions" as understood at the time. The French physicist and mathematician François Arago (1786–1853) offered an altogether fanciful account of a serendipitous stumbling on the problem. According to Arago (1833), in 1790 a physician had prescribed to a Bolognese woman afflicted with a cold a broth made from frogs. "Some of these animals," continued Arago, "already stripped by Mrs. Galvani's cook, were lying on a table when by chance an electrical machine at a distance from the table discharged its electricity. The [frog] muscles, though they were not directly stricken by the spark, experienced strong contractions." If this alleged event had set in motion the series of experiments recorded in Galvani's memoir of 1791, the pace of work would indeed have been extraordinary. The document *De viribus electricitatis in motu musculari commentarius* (Commentary Concerning the Effects of Electricity on Muscular Motion) had a modest fifty-five pages. It reported a key experiment conducted in 1786 that "verified that movements occurred in a dead and prepared [dissected] frog, without any external source of electricity, by applying a metallic electric arc solely on nerves and muscles" (Bernardi 2001). If electricity can elicit muscle contractions in a dead frog, then Galvani appeared justified in substituting a "neuro-electric force" for Haller's vague "irritability" or animal spirit in driving animal movements. The "demonstration" of intrinsic animal electricity created a furor among the savants of the late Enlightenment, and *De viribus* quickly became a bestseller. Such was the prevalent incredulity or skepticism that several naturalists busied themselves repeating Galvani's experiments on frogs. One of them was the physicist Alessandro Volta.

As a physicist, Volta stood in stark contrast to his fellow countryman Galvani, who had, in the words of Piccolino and Bresadola, a "unitary vision of natural phenomena" shaped by his training in philosophy and medicine. Born in Como near the border with Switzerland, Volta displayed a keen curiosity for physical phenomena even before adulthood (Arago 1833). By his twenties he had already produced two memoirs on electricity. Besides physics he made discoveries in chemistry, among which the swamp gas later identified as methane. In 1779 Volta was called to fill the chair of physics newly created at the University of Pavia. It was in this post that the echoes of Galvani's *De viribus* reached him. At first Volta thought highly of Galvani's experiments and, on repeating some of them, he seemed to agree with Galvani's interpretation of animal electricity. But Volta soon had second thoughts and took an antagonistic position.

In two letters to a fellow of the Royal Society, dated September and October of 1792 but published in 1793 (Volta 1816), Volta outlined at length not only his justification for rejecting the existence of intrinsic animal electricity, but also his clever experiments that allowed new insights into electrophysiology. As an experimentalist Volta had an advantage over Galvani in that he stressed the importance of quantifying observations and devising instruments to facilitate such quantifications. "What good can one expect, especially in physics," Volta argued, "if things are not reduced to degrees and measures? How can one evaluate the causes, if not only quality, but also the quantity and intensity of effects are not determined?" As Piccolino and Bresadola (2013) explained, Volta's attitude "suitably summarized the 'quantitative spirit' – the values of order, systematization, measurement, and calculation – that characterized scientific investigation in the latter part of the eighteenth century and distinguished it from traditional natural philosophy." Galvani, in contrast, leaned toward natural philosophy.

Two instruments built by Volta helped him in his investigations of "animal electricity": the electrophorus, which provided consistent and stable electrical charges for stimulating the frog neuro-muscular preparation; and the condensatore, which allowed measurements of the infinitesimal amounts of electricity sufficient to elicit muscle contractions (Piccolino and Bresadola 2013). These investigations led Volta to conclude that "the stimulating action of electricity acts primarily on the nerves ... while the motion of the dependent muscles is a secondary effect of nerves' excitement." This conclusion contradicted Galvani's assertion that muscle was directly excited by the "electrical arc," and Volta assumed that the threshold of electrical stimulation necessary to induce muscle contraction was much higher than that needed to excite the nerve contacting the muscle. Piccolino and Bresadola stress that Volta was on the right track: "[M]odern electrophysiology has proved that the electrical stimulus is more effective when applied to the nerves than to the muscles. Because of this higher nerve excitability, electricity acts primarily through the excitation of the nerve fibers present inside the muscle body, even when the stimulus is applied to the surface of a muscle."

To Volta's assertion that muscles have no intrinsic electricity, and that the disequilibrium of positive and negative charges between the outside and inside of the muscle, championed by Galvani, applies instead to the electric metal arc touching the muscle, Galvani responded with more experiments tailored to squash his rival's arguments. In a memoir addressed to Spallanzani, Galvani (1797) showed that, by making the sectioned end of a frog sciatic nerve touch the sciatic nerve on the other side of the body, contraction occurred on that other side and even sometimes on the side of the cut nerve. Piccolino and Bresadola again comment: "This experiment is generally considered a 'capital' experiment for the birth of electrophysiology. Indeed, neither Volta nor the other adversaries of animal electricity could propose any substantial objection capable of undermining the evidential value of the experiment as support of electricity intrinsic to the animal."

Beyond the adversarial character of the exchanges between these two great scientists, they shared one source of inspiration in their work: electric fishes. Swimmers are startled and pained by the electric shock suffered by contact with these fishes, but to Galvani that experience made palpable the notion of animal electricity at large. To Volta the structural organization of the fishes' electric organs served as a template for the design and construction of his *pile électrique* – the invention of the battery (Volta 1800).

Scientists of the late eighteenth century vaguely understood that animals could be useful in a variety of ways. They were capable of engineering feats worth emulating in search of practical solutions to problems of human daily life; and they presented ideal features for experimental work designed to address specific physiological problems. The two most famous Bolognese naturalists, Marcello Malpighi in the 1600s and Galvani in the 1700s, "were convinced that the investigation of [vital] functions in the animal body could offer the key to understanding the same functions in the human body, a view that gave a central role to comparative anatomy and physiology in the life sciences" (Piccolino and Bresadola 2013). The animal body most used for such a purpose at the time was the frog, if only because it offered, as Frederic Holmes (1993) put it, "a simplified, more accessible version of a mammal." Holmes even depicted the frog in the title of his essay as "the old martyr of science." All these and other uses of animals would later serve as justifications for a comparative approach to physiology.

Another useful approach to animal experimentation was comparing the functions of different animal groups to test how universal these functions are. Surprisingly, it was Volta the physicist, not Galvani the natural historian, who took this approach. In his second letter of 1792 to the Royal Society (Volta 1816), Volta described investigations of electrical stimulation on a variety of invertebrate animals:

> Experiment M. After sectioning the head of a fly, butterfly, beetle, etc. I slit their carapace with a penknife or small scissors; and I introduce deep in the slit, near the neck, a piece of a sheet of tin (the paper improperly called silvery is very appropriate) and a little over it I introduce the cutting edge of a silver blade, or of a small money coin: then when I move the latter in contact with the tin sheet, the legs begin to fold, to struggle, and the other parts, and even the trunk, to become agitated. I find it very amusing thereby to elicit the song of a cicada.

Volta concluded that only animals possessing definite limbs with articulation (flexors, extensors, and so forth) are susceptible to muscle contractions in response to electrical stimulation. As for those lacking such limbs, such as worms: "It is altogether a different animal economy, another type of mechanics for the movement of these animals." He added that only voluntary (skeletal) muscles are sensitive to the electrical effect. He was wrong; for all Volta's wizardry, the methods of his day lacked the sophistication to unveil what turned out to be the universality of the phenomenon of muscle sensitivity to electrical excitation.

The story of Galvani and Volta shows like no other that by the close of the eighteenth century the building blocks for the practice of comparative animal physiology were in place. But it was still an unspoken physiology, practised with little deliberate design. These haphazard ventures may not have created a subfield of physiology, but a shift did take place from a functional anatomist's practice of observing humans and familiar animal species to the preparation of rigorously designed experiments on animals of all descriptions. This shift allowed the possibility of new functional insights derived from the creation of innovative instruments. The soil was ripe for fertilization in the next century.

2

The Uncertain Beginnings

∾

Comparative physiology will be for us the science of life, considered as a
whole and in its particulars in all living beings, but especially in animals, that is,
the beings which live, feel and move according to Linnaeus's definition.
~ Antoine Dugès (1838)

The rise of experimental physiology coincided with the birth of biology. At
the turn of the nineteenth century, there was a growing dissatisfaction among
naturalists with the kind of natural philosophy epitomized by Galvani and
Volta, a mechanistic philosophy that subjected the study of living organisms
to the laws of physics or chemistry. Living processes were studied only by
analogy with physical processes. But to a new generation of naturalists it was
increasingly clear that the living world was more diverse and complex than
their predecessors had supposed.

The awakening to the variety and exoticism of the animal world was a long
historical process with gradual accretions. To put it in perspective, animal
domestication began about 10,000 years ago and private collections of exotic
animals began to make their appearance 5,000 years later (Kisling 2001).
These collections, held mostly by monarchs or wealthy families, appeared in
various civilizations around the world and grew in sophistication through
the ages. By the eighteenth century many such menageries, although private,
were to varying degrees made accessible to a gradually more enlightened
public. The transition from menagerie to zoological garden as a public institution with scientific aspirations is best represented by the Jardin des

Plantes in Paris, part of which was the Muséum d'histoire naturelle where the famous naturalist the Comte de Buffon worked. The spoils of the Napoleonic wars in Europe and Egypt supplemented the animal collections of the museum, and the resulting wealth of exotic specimens helped launch comparative approaches to animal studies in the first quarter of the nineteenth century. In parallel, the vast British Empire became a worldwide source for numerous zoological collections throughout the British Isles. The Hunterian Museum of the Royal College of Physicians in London (1799) and the Zoological Garden (1828) sponsored by the Zoological Society of London are prime examples of public institutions devoted to research, especially in comparative anatomy.

But there was more to the birth of biology than the study of diversity. Certain phenomena of life – parthenogenesis and tissue regeneration, for instance – are too peculiar and astonishing to be reduced to physical principles (Barsanti 1994). Living organisms, the new breed of naturalists argued, obey rules of nature uniquely their own. Gottfried Reinhold Treviranus (1776–1837) in Germany and Jean-Baptiste de Lamarck (1744–1829) in France – both credited with coining and disseminating the word "biologie" – proposed a program of study for the new field that departed from the constraints of mechanistic philosophy; they elaborated a theory of transmutation of species.

Treviranus, in the first volume of his book *Philosophy of Living Nature for Natural Scientists and Physicians* (1802), drafted his research program thus: "The objects of our investigations will comprise the different forms and manifestations of life, the conditions and laws under which the phenomenon of life occurs and the causes which determine it. The science which deals with these objects will be designated under the term biology or science of life" (Gayon 2005, based on a 1954 translation by Marc Klein). In the same year, Lamarck (1801–02), in his book on hydrogeology, separated the field of "terrestrial physics" into three theoretical domains: atmospheric (meteorological) theory, hydrogeological theory, and the theory of living bodies (biology). Further into the book, inspired by shell and coral deposits, Lamarck formulated what he claimed was "an accurate idea of the origin of living bodies, as of the causes for the gradual development and improvement of the organization of these bodies," adding that humans should be aware of the time and circumstances that had been necessary to arrive at the living species as they now exist. Both Treviranus and Lamarck observed that living beings obey basic rules that make life possible, but that they are also historical beings, organisms which, by their transmutations over time, generate a hierarchy of diversity of forms (Gayon 2005).

"It has passed unnoticed," remarked the French historian of physiology Joseph Schiller (1968), "that all through the nineteenth century biology and physiology developed independently of each other." In his view the outlook of biology, with its emphasis on the grand scheme of the living world and on species and their evolution, was at odds with physiology and "its analytical procedures, the use of the individual as working material, the deterministic explanation of each particular phenomenon excluding chance, conceived as an end in itself and devised as such by the experimental approach." As we will have occasion to appreciate later in this book, the outlook of biology, especially from the perspective of its subsections zoology and evolutionary biology, could have been reconciled with physiology through the channel of comparative physiology. But as textbooks of comparative physiology started to make their appearance in the first half of the nineteenth century, it became clear that such books were mere "descriptions of functions based on zoological classification" (Schiller 1968).

But before comparative physiology could begin to make its mark, according to Schiller, physiology in general struggled to do so at the dawn of the nineteenth century. Their battle for recognition came to the fore in France when Etienne Geoffroy Saint-Hilaire, a leading comparative anatomist and zoologist at the Muséum d'histoire naturelle, voiced concern that zoology and physiology were not fairly represented in the membership of the Académie des sciences (1822). Physiology, in fact, was not specifically named anywhere, let alone in the two sections of the academy where Geoffroy Saint-Hilaire thought it belonged: Anatomy and Zoology, or Medicine and Surgery. The latter section was peopled almost exclusively by undistinguished surgeons until one physiologist was elected in 1821, the very year Geoffroy Saint-Hilaire submitted his reproaching memoir (Crosland 1992). This physiologist was François Magendie (1783–1855), and to him devolved the merit of inaugurating the era of modern experimental physiology.

The first French physiologist of note at the turn of the century – an iconic product of the French Revolution – was Xavier Bichat (1771–1802), whose life of overwork and ensuing poor health was cut short at the age of thirty

(Shoja et al. 2008). After Robespierre's Reign of Terror ended in 1794, Bichat moved from his base in Lyon to Paris, where his fame – first as a surgeon, and later as a physiologist – grew rapidly at the Hôtel-Dieu Hospital. His reputation as an anatomical pathologist owed much to his book *Traité des membranes* [Treatise on Membranes] (1799). As a physiologist, Bichat is best known for his book *Recherches physiologiques sur la vie et la mort* [Physiological Researches on Life and Death] (1800), in which he classified life as either organic or animal. By organic life he meant the "vegetative" functions of internal organs; and by animal life the activities resulting from the organs – brain, sensory organs, and voluntary muscles interacting with an organism's environment (Pickstone 1981). In his physiological treatise Bichat crystallized his vitalism in this sentence: "Life is the sum of functions which resist death." His research method borrowed extensively from the "animated anatomy" of Hallerian physiology, and it was precisely this mold that Magendie was anxious to escape.

Magendie's desire to distance himself from Bichat was initially thwarted by the fascination that Bichat's organicist views exerted on the minds of *savants* and even artists in the early nineteenth century. A prime example from the literary world is the novelist Honoré de Balzac, who could be said to represent the generation of French writers whose curiosity was piqued by then current developments in biology to the point of incorporating them in their novels. Along with Goethe, Balzac was fascinated by the famous scientific debate of 1830 between Georges Cuvier and Geoffroy Saint-Hilaire over "whether animal structure ought to be explained primarily by reference to function or by morphological laws" (Appel 1987). Balzac had corresponded with the two rival biologists and in the debate he sided with Geoffroy's transcendental morphology. He even ridiculed Cuvier's stand on the primacy of function in a satirical piece of 1842 called *The Ass's Guide for the Use of Animals Who Wish to Achieve Honours* (Appel 1987).

Balzac was equally enthused, however, as Julia Przybos (2009) aptly explains, by Bichat's physiological outlook, which had already spawned a cultural progeny. Not least among those he had influenced was the epicurean Jean Anthelme Brillat-Savarin, in whose *Physiologie du Goût* (1826) one reads: "Death is the absolute interruption of sensual relations and the absolute annihilation of vital forces, which cede the body to the laws of decomposition." Przybos shows particularly how Balzac incorporated physiological concepts of digestion and sexual function into the narrative of some novels of his *La comédie humaine*.

Magendie followed a path similar to Bichat's in earning his medical degree and engaging in a career as hospital physician and surgeon (Flourens 1858). But his deep-seated interest in physiology found an outlet as early as 1821, when he initiated the *Journal de physiologie expérimentale et pathologique*, which provided a forum for his vision of physiological research. In 1831 a chair of experimental physiology was created for him at the Collège de France, which he held until his death in 1855. His contribution to physiology was no less than a seismic change of paradigm.

In a pamphlet published as early as 1809, Magendie expressed his disdain for Bichat's brand of vitalism, declaring himself cured of any penchant for theories or *esprit de système* (Albury 1974). Theories, he announced, were vain words, and in his *Journal de physiologie* he expressed this thought with a Cartesian quip: "You believe, therefore you don't know" (Dawson 1906). Yet for Magendie physiology as a discipline was in its infancy if only because it lacked the experimental framework that characterized physics and chemistry. Rather than deduce function from anatomical probing, which typified "the subservience of physiology to anatomy," as Schiller (1968) put it, Magendie proposed an approach that heralded "the proper dynamics of physiology": "In order for physiology to establish its supremacy, the trend has to be reversed: to start from the phenomenon, to follow its successive changes through the organism by way of experimentation, and to end with its localization. This step was taken by Magendie, who realized that a function transcends the activity of any single organ and conceived it as the activity of several organs converging as a system towards the same end."

Magendie made numerous advances in physiology: food swallowing and absorption, the role of the diaphragm in regurgitation and vomiting, and the role of the nervous system in the digestive process of birds and mammals (Magendie 1825). His pioneering study of the effects of drugs (emetin, strychnine) on animals led to the creation of the subfield of pharmacology. In these works the intellectual process that propels the experiments is quite transparent: initial probing leads to raising questions or feeding hypotheses, which are in turn answered or tested by designing a subsequent set of experiments, and so on. This process is precisely the way physiological experimentation is carried out even today.

A milestone of Magendie's physiology research consisted of mapping, by means of selective nerve cuttings, the sensory and motor divisions of the spinal roots to the limbs. This study led to a bitter struggle with the English surgeon and anatomist Charles Bell (1774–1842) for priority of discovery. As historian Gillian Rice (1987) has shown, Bell's approach to such a study, published in 1821, smacked more of "animated anatomy" than Magendie's new experimental spirit. Cleverly, as a good experimenter, Magendie took pains to select a simpler animal model for his purpose: a dog puppy. "In June 1822," Rice writes, "Magendie opened the unossified vertebral columns of live puppies to expose the posterior spinal nerve roots. First, he cut the posterior using small scissors, sutured the skin overlying the area and then observed the puppy." It was the first of several experiments that led to Magendie's correct mapping of the sensory and motor activities of the spinal cord. Bell, on the other hand, having seen from Magendie's work that he himself had reached some erroneous interpretations, republished his study in 1824, subtly altering it "to make his conclusions correspond with Magendie's findings and, at the same time, to give the impression that he had demonstrated and published these results by 1821" (Rice 1987). Had Bell been an academic scientist today, such scientific misconduct would have cost him his job and ended his career.

Magendie's experiments inadvertently launched the anti-vivisection movement, which has dogged animal experimenters to this day. Anita Guerrini (2016) has brilliantly shown how Magendie's work became the lightning rod of the anti-vivisectionists and how Bell's attack on Magendie, although not centred on the ethical issue of animal experimentation, served as an excuse for British Members of Parliament to debate whether to add vivisection to the Anti-Cruelty Law of 1822. A specific Vivisection Act eventually went on the books in 1876. However, the outrage against Magendie was focused on his demonstration during a visit to London in 1824 of an experiment on a dog which was clearly painful, "since no anesthetics were available and [Magendie] was unable to adopt Bell's practice of 'stunning' the animals since this would not have allowed Magendie to demonstrate the physiological foundation of sensory experience" (Guerrini 2016). The argument of serving the greater good of mankind evinced by the potential health benefits of experimentation would repeatedly be used to counteract anti-vivisection campaigns throughout the nineteenth and twentieth centuries.

Marie-Jean-Pierre Flourens (1794–1867), who wrote Magendie's eulogy (1858), followed in the latter's footsteps in short order. In fact, his name is often paired with Magendie's in singling out the two acknowledged founders of modern experimental physiology. Eleven years Magendie's junior, Flourens was a child prodigy who earned his medical degree in Montpellier at the age of nineteen (Vulpian 1888, Pearce 2009). Unwilling to practise medicine, he soon found his way to Paris, where he studied comparative anatomy and zoology under Cuvier and Geoffroy Saint-Hilaire, then eminent zoologists at the Muséum d'histoire naturelle. As Flourens mastered his field of choice, he made no secret of his burning ambition to become famous and vie for a seat at the French Academy, which indeed he won in 1840 (Vulpian 1888). While preparing a lecture on physiological theories of the senses, Flourens was intrigued by the questions the subject raised for experimental enquiry. Starting in 1821, he conducted numerous experiments, culminating in his monograph of 1824, *Recherches expérimentales sur les propriétés et les fonctions du système nerveux dans les animaux vertébrés.*

In the preface of this work, Flourens laid down the rules of experimentation that he followed and which his predecessors had failed to heed: choose animals of young age to minimize surgical and other complications and ensure long enough survival to observe results of interventions; change one experimental parameter at a time; limit ablations to the smallest circumscribable functional units. These measures were all designed to reach interpretations of results that were as unambiguous as possible. Through his experiments, Flourens sought to answer the following questions: Are sensory properties located in the same parts of the nervous system as motor properties? Are sensing and moving the same property in the nervous system or are they separate properties? Are the nervous organs of one property distinct from those of the other? His experiments led to the conclusion that: (1) sensory and motor functions are distinct in both nervous system location and effect; and (2) there are distinct boundaries that separate sensory and motor systems. One of Flourens's important contributions was to map the functional organization of the nervous system. He thus introduced the notion of command centres in the brain which organize activity in response to sensory input but do not directly excite muscles. He also defined the notion of coordination of motor and sensory systems to achieve, for example, a smooth locomotion or food manipulation, in which the cerebellum is involved.

Portrait of Marie-Jean-Pierre Flourens by Auguste Lemoine around 1860.
National Library of Medicine Image B07878.

Flourens valued natural history as much as medicine or physiology, and so it is not surprising that he did not limit himself to a single mammalian experimental model for his work on the nervous system as a medical physiologist would do, but branched out to representatives of every vertebrate type: mammals, birds, reptiles, amphibians, and fishes. He even tried his hand with squids. His goal was to find a unity of plan across vertebrate types in the functional organization of the brain and the peripheral nervous system. The comparative approach necessary for his search for a unity of plan may have been what inspired him to produce three decades later his course in comparative physiology at the Muséum d'histoire naturelle (Flourens 1856). But, curiously, a perusal of those lectures reveals that his understanding of comparative physiology was greatly at variance with the mission of the field as it is understood today. To Flourens comparative physiology meant that the individual organism and the species were the centre of interest, not the study of body parts. In fact, the greater part of his course deals with the development of animals from the egg and with paleontology. This stance, including the view that species are fixed and cannot be transmuted (or evolve), reflected the influence of his old mentor, Georges Cuvier.

∼

Flourens's inchoate attempt to find a niche for comparative physiology was not the first. Frédéric Cuvier (1773–1838) worked for decades in the shadows of his older brother, Georges, also as comparative anatomist and paleontologist, only to become the recipient in 1837 of a newly created chair of comparative physiology at the Muséum d'histoire naturelle a year before his death (Flourens 1840). The appointment meant nothing for the advancement of comparative physiology and smacked of being politically motivated. To read what seems to be the first rallying cry for the creation of the field of comparative physiology, we can hark back to the year 1825. In a book by Julien Joseph Virey (1775–1844), a naturalist and anthropologist better known for his racist views and vitalist principles than for any rigorous scientific thinking (Corsi 1987), we find this passage:

Any hypothesis that does not embrace the universality of life phenomena in all living organisms, from man to the polyp, and from the cedar to the lesser lichen, cannot represent what goes on in nature, or could not reflect the true state of things. For this reason we need a *comparative physiology* in the same way that we currently have a comparative anatomy. (Virey 1825)

A mere five years after Virey's entreaty, the first book entirely devoted to comparative physiology appeared. *Principes de physiologie comparée* (1830), a volume of more than six hundred pages, was authored by Isidore Bourdon (1796–1861). What little we know of Bourdon is distilled in anonymously published obituaries of the *Bulletin de l'académie royale de médecine* (Vol. 27: 160–2, 1861) and *Le moniteur scientifique* (Vol. 4: 60–1, 1861). At first a student of Georges Cuvier, Bourdon quickly switched to medicine, and his work in medical physiology soon earned him membership in the Royal Academy of Medicine. He also served in official missions during cholera epidemics and as doctor in charge of epidemics in the Seine Department. He was medical inspector of mineral waters for five years. He was said to possess a sharp pen but knew how to turn a phrase so as to severely criticize an author without hurting his feelings.

Bourdon was only thirty-four when he published his textbook of comparative physiology, which came on the heels of his *Physiologie médicale* (1828). He had envisaged two books to encompass the various aspects of comparative physiology, but only the first found its way into print. This book was divided into four parts: general physiology, and comparative aspects of reproduction, growth, and nutrition. In its scope, the book largely anticipated Flourens's approach. Had it been published, the second book, announced in the preface of the first, would have explored animal functions more in line with the future development of the field: respiratory functions, thermoregulation, secretions, the circulatory system, and the nervous system. While the intended coverage of material met the canons of the burgeoning field, Bourdon was silent on the *raison d'être* and goals of comparative physiology itself, without which it could hardly be elevated to the rank of scientific discipline. His approach was descriptive, with little attention to mechanisms.

The next book whose title page contains the words *physiologie comparée* was written by Henri Ducrotay de Blainville (1777–1850). Blainville was a zoologist and anatomist in the tradition of his teacher Cuvier, with whom he soon came into conflict owing to his difficult or "contrary" character

(Flourens 1854). He had little or no training in physiology and yet he taught "comparative physiology" at the Collège de France, and his book was a compendium of his lectures (Ducrotay de Blainville 1833). In comparison with Bourdon's *Principes de physiologie comparée*, Blainville's book introduction includes a prolix and somewhat confusing discourse on the position of comparative physiology within the biological sciences. Blainville divided zoology into six branches, one of which, zoobiology, includes – or is synonymous with – comparative physiology. By zoobiology he meant "the science which analyses in animals the phenomena of life through their production, their relationship with levels of organization or with the outside world, and which seeks to explain these phenomena by linking them to general laws of matter whenever applicable." His definition does not stray far from the view that physiology as a field of study is interchangeable with natural philosophy and its emphasis on the idea that the dynamics of life phenomena obey laws of physics and chemistry (Hagner 2003). For Blainville, to study comparative physiology was to search for evidence of the unity of life and matter, as basic components of life (gases, minerals, etc.) are derived from, and interchangeable with, materials in the inanimate world surrounding living organisms.

Not only did Blainville construct a philosophical justification for comparative physiology but he also proposed a program of study for the embryonic field, about whose future he was less than optimistic. He laid down a detailed methodological protocol for achieving results: (1) direct observation of the various parameters of animal activities, in their spontaneity and in response to outside interventions; (2) comparison of the circumstances associated with these animal activities according to age and the different animal groups in the "scale of organisms" to deduce which life phenomena are universal or commonly shared; (3) study of the contribution of abnormal organizations or pathologies to the understanding of normal functions; (4) experimentation to answer specific questions; and (5) examination of brain mechanisms by which relations of cause and effect in observed phenomena can be seen and distinguished.

A fitting example of Blainville's comparative approach is found in his treatment of blood in his tenth lecture. He observed that blood volume, even when corrected for body size, is consistently larger in vertebrates than in invertebrates (this statement is no longer considered valid). Blood is red in colour in all vertebrates whereas it is colourless in many invertebrates. He found that "in birds, who are highly efficient breathers, the distinction between arterial and venous blood is more clear-cut than in mammals." The relative venous volume, he claimed, is larger in aquatic mammals such as seals and whales than in terrestrial mammals. In invertebrates it is often difficult, if not pointless, to discriminate between "venous" and "arterial" blood in the traditional sense. He found no clear trend in the animal series for the size and number of blood cells. These vague statements reflect the state of the science and the methodological limitations of the period, when the quest for underlying mechanisms of animal functions was neglected. Both Blainville, who had ambitious but thwarted hopes for the field, and Bourdon, misguided in his approach and losing interest in his book, contributed to the difficult birth of comparative physiology.

A contemporary who acknowledged the difficulties and setbacks of comparative physiology as it struggled for emergence was Antoine Louis Dugès (1797–1838). Born in Charleville (Ardennes) – a birthplace he shared with Arthur Rimbaud – Dugès obtained his medical degree in Paris in the early 1820s with a specialization in obstetrics. At the age of twenty-six he was appointed professor of obstetrics and eventually dean at the Montpellier Medical School (Dechambre 1884). Dugès dated his interest in natural history and physiology back to his move to Montpellier in the south of France, a region that displayed an enticing and vibrant diversity of living forms. Dugès's investigations there, conducted alongside his medical duties, led to an ambitious three-volume treatise of comparative physiology (Dugès 1838). His untimely death in 1838 prevented him from seeing the publication of his books, a task that was completed by friends and colleagues.

In the introduction to the first volume of his *Traité de physiologie comparée de l'homme et des animaux*, Dugès explained that his goal was to provide a comprehensive and balanced treatment of comparative physiology for the instruction of zoology students. This statement was an implicit recognition that comparative physiology should be considered a branch of zoology, as Blainville had insinuated five years earlier. "Although founded entirely on analytical studies," Dugès wrote, "that is, moving from the particular to the general, our physiology, like all constructed science, presents topics in a synthetic form." With this phrasing Dugès appears to push for the elevation of comparative physiology to the rank of a legitimate scientific discipline. In this respect, he was the first to support his narration with numerous drawings of a didactic nature and copious citations from the scientific literature.

Dugès stated clearly what he meant by animal functions: "By this word we designate any activity (simple or elementary functions) or any series of activities (complex functions) performed by living organisms which tend (for the latter) toward a common goal and are useful to the individual." He stressed the importance of the coordination of activities in complex functions, citing as an example the coordination of several muscles by the nervous system to execute a movement or food manipulation task.

At another level, Dugès compared functional systems from the lowest forms to mammals, in an attempt to show a gradation in complexity as one moves up the scale of animal types. As Darwin's *Origin of Species* was still twenty years in the future, no talk of evolutionary trends was attached to the interpretation of these comparisons. Dugès paid great attention to the nervous system and the sensory functions. He was the first to describe in detail the nervous systems of invertebrates and to show the great extent to which their functional organizations differ from those of vertebrates. He anticipated the notions today called habituation (*habitude*) and facilitation (*surexcitabilité*), which he rightly associated with learning. These processes became understood at the cellular level only in the 1970s, using a mollusk model, the sea slug *Aplysia* (Castellucci et al. 1970; see chapter 9).

Dugès exhaustively covered classical topics such as thermoregulation, respiration, circulation, and nutrition. These and muscle contraction and neuro-muscular action tend to be universally present across the range of animals. But he went further, discussing functions such as electric discharge by fishes, luminous organisms (firefly), and colour changes (chameleon, squids), which are the preserve of a limited range of animals, and he made these functions part and parcel of comparative physiology.

In his discussion of the reproductive system, Dugès treated at length the egg and its development, and touched on a debate that had raged since the previous century. The debate separated those who believed "that complexity [of forms] must be imposed from without by some vital force ... working upon an egg that had only the potential for normal development (epigenesis)" from those who favoured "the notion that all major structures of the adult are already preformed in the sex cells (preformation)" (Gould 1977).

Strangely, Dugès and his contemporaries used the word "evolution" to signify preformation. Today it seems counterintuitive that preformed structures would evolve. Be that as it may, Dugès produced numerous arguments and observations to dismiss preformation. The use of the word "evolution" in its modern sense only appeared another thirteen years later, when the British philosopher Herbert Spencer (1852) wrote a short essay about "the Theory of Evolution," whereby new species would be produced by "continual modifications due to change of circumstances."

~

In the mid-nineteenth century, under Henri Milne-Edwards and Jean Louis Armand de Quatrefages, French animal physiology moved away from the tradition represented by Dugès. These two influential figures epitomized a new brand of zoologist who not only devoted their entire careers to researching various invertebrates from anatomical and physiological perspectives but also conducted their work in the field, thus gaining unique insights into how these animals function in their natural environment. Although they followed in the tradition of Cuvier at the Muséum, their paths diverged greatly from their predecessor in that their outlooks on functional systems and physiological mechanisms portended a new zoological project. The title of the series of volumes by Milne-Edwards on the subject – *Leçons sur LA PHYSIOLOGIE* [book's capitals] *et l'anatomie comparée de l'homme et des animaux* – seemed, for instance, designed to emphasize comparative physiology rather than comparative anatomy.

The story of Milne-Edwards is so rich in anecdotes and accomplishments that only a substantive biographical introduction will do justice to the man and his scientific opus. Henri Milne-Edwards was born in Bruges in 1800 at a time when the Belgian city was part of France (Berthelot 1891). This circumstance made it easier for him to later claim French citizenship, even though Bruges had reverted to Belgium. His father, William Edwards, owned a plantation in Jamaica and had returned to live first in England, and then in Belgium when Henri was born. Henri's mother, a French citizen, was William's second wife and Henri was the twenty-fourth child of William's remarkably large progeny. It is a strange coincidence that two great French scientists of the nineteenth century were both products of a mixed marriage

Portrait of Henri Milne-Edwards by Eugène Pirou in 1883.
Bibliothèque nationale de France.

between an English-speaking father and a French mother. The famous medical physiologist Charles-Edouard Brown-Séquard (1817–1894), born in Mauritius and destined to become professor at the Collège de France (Aminoff 2011), shared this parental history with Milne-Edwards.

When Henri was seven, his father was imprisoned by Napoleon's police for having facilitated the escape of British prisoners in Bruges. In 1814, by which time Henri was a teenager, his father was freed and reunited with his family (Berthelot 1891). The family took up residence in Paris, where they lived comfortably. Henri studied medicine and lived the life of a dilettante, seeking the company of fellow medical students and artists associated with the Sorbonne. In 1822 he became acquainted with the young Henri Beyle – the future Stendhal – who was a friend of Henri's brother Edouard. Henri's early interest in natural history as well as medicine is borne out in Stendhal's memoirs, *Souvenirs d'Egotisme* (revised edition of 1927), in which the famous novelist colourfully describes Henri's activities:

> He killed one thousand frogs a month and was on the verge, it is said, of discovering how we breathe and a treatment for the maladies de poitrine [tuberculosis, pneumonia] of pretty women. You may know that when leaving a ball the cold air kills each year in Paris one thousand and one hundred young women. I have seen the official statistics.

The humorous tone conceals the fact that Henri was investigating the microscopic structure of tissues across vertebrates – including frogs – and the results, purporting to show that the constituent cellular elements are similar in size and appearance in all vertebrates, formed the corpus of his first published article (Milne-Edwards 1823). He foreshadowed the cellular theory of Theodor Schwann and Matthias Schleiden, but the poor quality of his microscope caused him to miss the diversity of the shapes and sizes of his "globules," later known as cells.

The carefree life of Milne-Edwards ended in 1825 when the family fortune vanished, and Henri had to fend for himself and his young spouse. He resorted to friends and various expedients to survive financially. But this setback did not deter him from launching his research career in zoology, and as soon as 1826 he and his friend Jean Victoire Audouin (1797–1841) were busy investigating the circulatory system of crabs on the Normandy coast.

Until that time, investigations of animals' functional systems had been based largely on preserved museum specimens and involved deducing function from anatomical organization, as Georges Cuvier had prescribed. These investigations were predicated on the predictive value of known functions of an organ in one animal group for other groups in which organs presented similar organizations. This method worked fairly well for vertebrates, but invertebrates display such bewildering diversity that this approach, in Milne-Edwards's mind, was doomed to failure. Only direct observations and experimentation on live specimens would satisfy the zoologist's quest for understanding the inner workings of a variety of animals. This was the research program initially conceived by Milne-Edwards, a program that contained the elements of comparative physiology without spelling them out.

Audouin and Milne-Edwards (1827) demonstrated by careful perfusions of traceable dyes that in crustaceans, blood (presumably oxygenated) circulates from the heart through arteries to irrigate the organs. From these organs blood flows slowly to large venous sinuses from which vessels carry deoxygenated blood to the gills. Another series of vessels on the opposite side of the gills bring the just-oxygenated blood back to the heart. In support, Audouin and Milne-Edwards provided exquisitely detailed illustrations in which the crustacean circulatory system was clearly mapped. The work was the foundation stone of what became known as Milne-Edwards's school of "physiological zoology."

To feed his growing family, Milne-Edwards took teaching jobs that limited the time he could spend on research. For a while he taught at the prestigious Lycée Henri IV in Paris, but in short order he landed a professorship at the Ecole centrale des arts et manufactures, where he taught natural history and hygiene (Berthelot 1891). Presumably to compensate for the limitations on his investigations, he took to reflecting on the significance of his research so far, and as soon as he had completed his work on crustaceans Milne-Edwards started developing his theory of the "division of physiological labour," which was to consolidate his fame in years to come. His first formulation of the idea served as the conclusion to an entry in the encyclopaedic *Dictionnaire classique d'histoire naturelle*, in which he followed the increasing levels of organizational complexity in the nervous system, from the lowest animal forms to the highest (Milne-Edwards 1827a). He presented his thinking in simple words:

It is first the same organ which feels, moves, breathes, absorbs nutrients around it, and ensures the preservation of the species; but little by little these various functions have each their own instrument, and the various actions inherent to these functions are executed in a distinct organ. Nature, ever economical in the means by which she reaches her goals, has followed for the improvement of living beings the principle so well developed by modern economists, and it is in her works as well as in art creations that one sees the enormous advantages resulting from the division of labour.

In another entry in the same encyclopaedia, Milne-Edwards (1827b) expanded on this concept. To make his point he offered the striking example of nutrition within cnidarians. In hydra nearly all epithelial cells are involved in absorbing and processing nutrients from their immediate surroundings, so that there seems to be no functional specialization. Cutting hydras in many pieces does not endanger survival, as the epithelial cells in the cuttings continue feeding and a new hydra is reconstituted by regeneration. In contrast, another cnidarian, the jellyfish, is of such a size as to make the hydra's feeding method unworkable. As a result the jellyfish had to develop specialized organs and compartmentalize the feeding function; hence the division of labour.

As historian of science Camille Limoges (1994) explains, it is not clear whether Milne-Edwards's concept, explicitly called "division of physiological labour" in his lecture notes of 1831–32 (Milne-Edwards 1834), was borrowed from Adam Smith's notion of division of labour for economic productivity in *An Inquiry into the Nature and Causes of the Wealth of Nations* (1776), or whether Milne-Edwards "developed these views at greater length in an environment particularly congenial to associations between natural history and political economy, the Ecole centrale des Arts et Manufactures, where he started to teach in 1831." Limoges sees an analogy between Smith's and Milne-Edwards's concepts of the division of labour:

Division of labour, according to Adam Smith "increases the productive powers of labour," it "increases the dexterity of the workers," saves time, increases the quantity of work an individual is able to perform, and favors the progressive invention of new means to facilitate and

abridge work. According to Milne-Edwards, the benefits of the division
of physiological labour roughly parallel those of the process applied
in modem manufactures: it increases the "vital powers" of the organism, improves the "quantity [grandeur] of the results" and the "quality
of the products" of "vital work"; moreover, once started, the tendency
of nature is to "increase the number of dissimilar parts and the complication of the machine."

Limoges also masterfully shows how Darwin, who admired Milne-
Edwards's work, borrowed from the Frenchman's concept of the division of
physiological labour to account for cases of divergence in phylogenetic trees.
In Darwin's view, divergence could be explained by a process of "division of
ecological labour," whereby animal groups occupy increasingly specialized
ecological niches.

The concept of division of physiological labour became a central tenet
of Milne-Edwards's physiological zoology, as to him it best accounted for his
large-scale observations on a variety of animals: cnidarians, crustaceans,
insects, and tunicates. Historically it was perhaps the first time that a comparative physiological approach had yielded a theoretical principle of wide
import. One of the goals of comparative physiology that was to be articulated
in the next century – to find common physiological principles out of the
diversity of animal forms – had thus been reached early in the development
of the field. However, as Limoges argued, Milne-Edwards's principle, although alluring, lacked the kind of biological foundation that Darwin built
for his evolutionary principle, and as a result the Frenchman's concept of
division of labour did not outlive the nineteenth century.

Now, while Milne-Edwards actively promoted physiological zoology, he
could not conceive of physiology without anatomy. This view permeated his
monumental treatise, *Leçons sur la* LA PHYSIOLOGIE *et l'anatomie comparée
de l'homme et des animaux*, which ran to fourteen thick volumes and spanned
the years 1857 to 1881. By then Milne-Edwards, highly reputed for his solid
achievements, had moved on to a prestigious professorship at the Muséum
d'histoire naturelle (1841) and had been elected dean of the Faculté des sciences de Paris.

If physiology and anatomy are inseparable aspects of one science, as
Milne-Edwards asserted in the introduction to the *Leçons*, then physiological

activity could be seen as the determining force shaping the organism: "There is always harmony between functions and organs; but what dominates in the living being and somehow dictates its proper nature is the manner by which the forces at play must exert themselves in its organism, and not the manner by which its organs are constituted" (Milne-Edwards 1857). What is equally striking in the *Leçons* is the tension between diversity and commonality as emerging principles for reviewing functions across the animal world. At times one can be mesmerized by the diversity among species of solutions to a functional problem, but one can also see that disparate species arrived at common solutions. This tension is perfectly captured in the following passage:

But when one studies attentively the entire animal kingdom, one quickly realizes that Nature, while largely obeying the law of the diversity of organisms, obeys also a *law of economy*. She has not put to use all the physiological possibilities, and shows herself less enticed to innovate if the innovations grow in importance. It seems also that before resorting to new resources to diversify its products, she wished first to try each of the processes put to use to obtain dissimilarities, and as much as she shows herself prodigal of varieties in the works of Creation, she seems inasmuch economical in the means by which this richness of results is attained. (Milne-Edwards 1857)

Milne-Edwards's *Leçons* dealt with all aspects of what was later labelled "environmental and metabolic physiology": blood characteristics, respiration and respiratory organs, blood circulation, heart function, circulatory system, food absorption, digestive system, secretions, urine production and excretion, heat production and thermoregulation, and metabolism. The latter word was introduced by Theodor Schwann in 1839 – in his epochal book delineating the cell theory – to designate chemical processes in the cell (metabolic phenomena). More than twenty years after the word was coined, Milne-Edwards seemed unaware of the word and instead wrote of oxidation and combustion in general terms. Integrative physiology – reproduction, sensory and motor neurophysiology – was also covered, and he concluded the treatise with a look at the division of labour in animal societies and social behaviour at large.

If Milne-Edwards embraced the comparative approach, which allowed him to see the functional changes that occur among animal groups, the *Leçons* and other works betrayed his old-fashioned view of Nature as a moral determinant or deistic substitute that shaped these changes in the background. The arrival on the scene of Darwin's *Origin of Species* just as the volumes of the *Leçons* were coming out soon made Milne-Edwards's interpretations of functional modifications obsolete. By the time the all-consuming *Leçons* were completed in 1881, their author only had four more years to live.

Not only did Milne-Edwards leave a considerable intellectual legacy in his physiological zoology but he also trained or influenced many French zoologists who enjoyed successful careers. Among those who attended the celebration of the completion of the old professor's *Leçons* was his own son Alphonse (1835–1900), who followed in his father's academic footsteps and became a distinguished classical zoologist based at the Muséum d'histoire naturelle in Paris. Another was Jean Louis Armand de Quatrefages de Bréau (1810–1892). I have touched in a previous book on Quatrefages's accomplishments as a researcher of marine bioluminescence (Anctil 2018), but as one of Milne-Edwards's keen disciples he became also a practitioner of physiological zoology.

Quatrefages earned his first doctorate, in mathematics, in Strasbourg before his twentieth birthday (Hamy 1894), but his interests soon shifted to medicine. After completing a medical thesis in 1832, he practised that profession in Toulouse, becoming absorbed in addition by zoological research on the side. Frustrated by the limited horizon of natural history in provincial France, Quatrefages moved to Paris, where, in 1840, he obtained yet another doctorate at the Muséum. The following year he met Milne-Edwards, then the newly installed Muséum professor, who convinced the talented Quatrefages to study marine invertebrates (Hamy 1894). In 1844 Quatrefages accompanied Milne-Edwards and another disciple in a daring (for the time) expedition to study the invertebrates of coastal Sicily. Quatrefages, in his *Souvenirs d'un naturaliste* (1854), recalled how Milne-Edwards used that occasion to try a *scaphandre* (diving suit). Posterity would attest that he could boast to be the first scientist in history to do so. Quatrefages's description gives a sense of the makeshift nature of the contrivance available at the time:

The apparatus used by Mr. Edwards for these submarine walks was one invented by Colonel Paulin, former chief of Paris's firemen, to combat basement fires. A metallic helmet with a glass visor covered the head of the diver and was secured to the neck with a leather apron fastened in place with a padded collar. This helmet, truly a miniature diving bell, communicated by a flexible tube with an air pump handled by two of our men; two others were kept in reserve to replace the first two. The remaining crew held the end of a rope which, running through a pulley attached to the boat yard, was fastened to a kind of harness and allowed a rapid rise [two minutes] to the surface of the diver whose heavy lead soles had pulled to the bottom.

Thus rummaging around on the seabed, down to a depth of eight metres, Milne-Edwards observed animals in their natural environment and could hand-pick specimens. A unique usefulness of the diving apparatus was to allow its wearer to retrieve eggs of mollusks and annelid worms that he could not have gathered by other means. In this way he and Quatrefages greatly enriched the field of comparative embryology, little known in those years. Understandably Quatrefages was in awe of Milne-Edwards, who thought little of risking his life for the sake of zoology. He emulated his older colleague by embracing fieldwork in maritime sites throughout his career.

The contribution of Quatrefages to physiological zoology is modest compared to that of his mentor and, for that matter, smaller than to his own purely zoological and anthropological studies. Nevertheless, he produced a multi-volume monograph on annelid worms, in which a small part dealt with the digestive, respiratory, circulatory, and other functions (Quatrefages 1865). He made physiological observations on a variety of bioluminescent animals (see Anctil 2018); and he made a comparative study of the physiological events associated with development from egg to adulthood throughout the animal kingdom (Quatrefages 1862). Quatrefages corresponded with Darwin but opposed the theory of natural selection for lack of evidence (Sillard 1979). He was persuaded that transformations did occur in the history of the animal kingdom, but he refused to endorse a structure for these transformations if unsupported by solid observations. Like Milne-Edwards, he put great store in the moral and spiritual character of humans, and in his book *L'espèce humaine* (1877) he placed humans in a separate kingdom from other primates on that basis.

French scholars such as Quatrefages considered themselves natural historians first, and physiological inquiry was either an occasional extension of their zoology or was embedded in their vision of what zoology stood for. The process that produced the generation of zoologists fully dedicated to physiological pursuits was yet to come, as the upcoming chapters will show.

3

Building Sound Foundations

∿

In general, I would point out that there are two ways open for us to obtain
knowledge of nature's secrets. We may merely watch natural events as
they occur or we may arrange conditions so that the events will appear,
disappear, or be modified as we may decide.
~ Walter Bradford Cannon (1945)

While French physiological zoology underwent halting progress in the early
and mid-nineteenth century, German physiologists were less successful in
contributing to the emergence of the field. It seems that they for the most
part only paid lip service to animal physiology, using animals such as frogs
to ask fundamental physiological questions, not to seek out physiological
principles by comparing animal species. But the focus of their concentration
was nevertheless of utmost importance for the future of animal physiology
in that they – along with one outstanding French physiologist – helped to
frame the foundational principles of modern physiology.

The originator of a whole line of such German physiologists was Johannes
Peter Müller (1801–1858). (Laura Otis's insightful book on Müller's life and
his students' perceptions of him [Otis 2007] serves largely as the basis for
these biographical notes.) Johannes Müller was born in Koblenz in the
Rhineland, a part of Germany then occupied by Napoleonic forces, just a
year after Milne-Edwards's birth. (Their contemporariness is only one thing
they shared, as we shall see.) His father was a cobbler whose meagre resources
were seriously drained by giving his son a good education. Nevertheless,

Müller attended the local *Gymnasium* to prepare for university study. In the first of many such actions by patrons who detected and nurtured Müller's gifts, a Koblenz educator convinced the boy's father that his son would fulfill his destiny better by attending university than by apprenticing as a shoemaker. Thus Müller embarked on the study of medicine at the University of Bonn, where he graduated in 1822.

The curator of Bonn University secured a scholarship for Müller to continue his studies in Berlin under Karl Asmund Rudolphi (1771–1832), the professor of anatomy and physiology who had founded the Berlin Museum of Natural History in 1810 and pioneered the field of parasitology. Under Rudolphi, Müller became skilled in the use of the microscope, and that instrument became a constant research companion. His early work showed much originality. When he was barely twenty-two, he examined the mammalian foetus by opening up the wombs of pregnant animals and studying how the foetus's separate blood supply provides oxygenation (Müller 1824). His judicious experimentation not only gave him information about the living conditions of the foetus, but also contributed to his rising reputation in his alma mater, Bonn, where he was hired as lecturer in anatomy and physiology in 1824.

From the standpoint of comparative physiology, Müller's years of lecturing in Bonn (1824–32) were of critical importance, as they resulted in landmark studies of the visual, circulatory, endocrine, and reproductive systems. "He rarely studied any function in one animal alone," Otis (2007) remarked, "preferring to compare the ways that different organisms solved physiological problems." A case in point is Müller's epochal monograph of 1826 on the visual functions of men and animals. In it he contrasted the very divergent ways in which the eyes of vertebrates and insects are organized and process visual information. But in the same work he also derived a unifying principle about the way the external world is perceived by our senses, based largely on self-experiment. He called it the *law of specific energies of the senses*.

By this law Müller posited that each sensory organ produces its own sensory energy irrespective of the nature of the external stimulus. For example, with one's finger one can put pressure on one's skin and produce a sensation of touch, but similar finger pressure on the eyeball can also produce a sensation of light. If one can perceive light by pressing on an eyeball through closed eyelids, then, he concluded, our perceptions do not give us a true im-

Portrait of Johannes Peter Müller.
National Library of Medicine Image B19893.

pression of the physical features of the world around us. This was no inno-
cent statement on Müller's part; it implied that our senses cannot be trusted
to render an objective image of our surroundings. As Walther Riese and
George Arrington (1963) explained: "We believe that the most far-reaching
effect of the law of the specific energies of the senses on the structure of
medical thought is to be seen in the shifting emphasis from objective data

to subjective experiences. This was a break with the nineteenth century concept of science and unquestionably was rooted in Muller's vitalistic and animistic heritage."

Müller's vitalism is also emphasized by Otis (2007): "He believed that nature's diverse forms had arisen as a result of the unique structure of living matter. In this respect, he can be considered a vitalist, but his belief in 'life forces' drove him to conduct experiments. As a physiologist, he supported any study that might show how organic matter gave rise to life functions." Another aspect of the approach to physiology that Müller expounded in his *Visual Sense* monograph was his insistence that observation of the activity of an organ leads to a better understanding of function than experimentation. Several of his own students, as Otis documents, believed that experiments were paramount and they spent their careers contradicting their mentor's dictum. But the entire debate was based on a misunderstanding. It is one (albeit worthwhile) thing to observe a phenomenon and quite another to understand the underlying mechanism of the phenomenon, which can hardly be achieved without experiment. Müller believed that observing a life phenomenon raises questions about its causes, questions that can then be answered through experiment.

In the years following his opus on the physiology of sight, among his other endeavours Müller looked for analogs of the vertebrate sympathetic nervous system in insects (Müller 1828), and he discovered the physiological mechanism (nervous and circulatory) underlying male erection in mammals and birds (Müller 1835). In recognition of his growing fame, and again with help from his patrons, Müller was offered the chair of anatomy and physiology at the University of Berlin. Securing the chair was far from a done deal, however, as politicking and jostling among colleagues in Berlin eager to push forward their own applicants made the outcome uncertain. Müller had to plead his case to government officials. In a letter to Prussian culture minister Karl vom Stein zum Altenstein, he "emphasized the importance of physiological research for medical advancement and echoed Humboldt's desire to make Berlin a research center equal to Paris" (Otis 2007). Here Müller was probably referring to the kind of physiology professed by the likes of Milne-Edwards at the Muséum d'histoire naturelle.

In the end Müller was persuasive but, as he settled in Berlin in 1833, he increasingly turned away from physiology – comparative or otherwise – and immersed himself in comparative anatomical studies and the collection of field specimens for his museum. As Otis remarked: "[H]e was always in close contact with his research animals in their native environments. While his microscope revealed life at the cellular level, it also kept him in touch with living organisms." This dual interest he shared with his French counterpart Milne-Edwards. Although shying away from original physiological research, he produced his *Handbuch der Physiologie des Menschen für Vorlesungen* (1837–40), a highly influential text that defined the field and earned him the accolades of colleagues across Europe as the foremost physiologist of his era.

As celebrated as Müller was, his rewards were partly tarnished by the personal issues that dogged him for most of his adult life. Once he had decided on a life of scholarship and scientific research, he developed a cyclical pattern of restlessness, workaholism, and insomnia followed by collapse in the form of a nervous breakdown or burnout. This pattern not only put him out of commission for weeks or months on end but also put a strain on his family. As Otis (2007) observed, "Müller's workaholism, anxiety, insomnia, and loneliness acted synergistically, so that as he aged and his best-loved students left him, he grew increasingly depressed." Add to that the difficulty of facing radical students' demands for reform in his role as rector of Berlin University during the Revolution of 1848, and the enduring trauma of his near-death experience in 1855 when the ship he was travelling on for a field expedition was sunk in a storm, and you have the recipe for the depression spiral that culminated in his untimely death at the age of fifty-six.

In his Berlin years, Müller attracted many students who helped establish the University of Berlin as the best centre of attraction for physiological research in Europe and, for that matter, the world. If Müller's path diverged from physiological work, it was not for lack of understanding the importance of physiology for the future, as several of his students were encouraged to pursue physiological research. Standing out from the others are three prominent physiologists whose contributions are relevant to the topic of experimental and comparative physiology: Emil du Bois-Reymond (1818–1896), Ernst Brücke (1819–1892), and Hermann von Helmholtz (1821–1894). A fourth, Carl Ludwig (1816–1895), came from outside the Berlin circle of Müller's disciples.

All four, from the very beginning of their careers, formed a tightly knit academic brotherhood.

If these four stand out, it is less for their contribution to animal physiology proper – their outlook was largely paramedical – than for embracing an experimental physiology based on the design of sensors and recording instruments and for methodologies that established physiology on a firm physico-chemical basis and pushed out the vitalist outlook preponderant among their predecessors. Just as the 1848 Revolution erupted, du Bois-Reymond articulated their "manifesto" for a new physiology in the preface of his newly published *Untersuchungen über thierische Elektricität*. He declared that the "path of destiny" for physiology to follow was to discard the "life force" (vitalist) approach and accept that all life processes can be reduced to physico-chemical phenomena and investigated as such. In this context he and his co-signatories labelled themselves "organic physicists," thus emphasizing their reductionist philosophy. Not only did the rapid advancement of their research hurtle physiology into the modern era but their instrumentation and their methods eventually became the standards for comparative animal physiology. For these reasons it is important to examine carefully the achievements and philosophy of the four scientists in question.

Du Bois-Reymond's biographer, Gabriel Finkelstein (2013), rightly points out that he was "the most important forgotten intellectual of the nineteenth century." In his day, du Bois-Reymond was renowned both as an outstanding physiologist and as a popular essayist on various philosophical and historical topics whether or not relevant to science. Born in Berlin to a Huguenot family, he was raised speaking French while also learning German. This *émigré* background left him feeling like an outsider much of his life. Attracted first to the arts, he soon found his calling in science as he toggled between the University of Berlin and Bonn University, where he focused on physics and chemistry. But, as Finkelstein takes pain to emphasize, du Bois-Reymond's newly found attachment to these sciences did not sway him from the "disease" of the period: an infatuation with *Naturphilosophie*. This philosophy of nature, born at Jena University of the Romantic philosophy championed by Friedrich Wilhelm Joseph Schelling (1775–1854) among others, was a theoretical construct emphasizing the unity of living beings as an integrated whole. It culminated in the "philosophical anatomy" of Etienne Geoffroy Saint-Hilaire, which claimed that all vertebrate animals share a single body plan. As Toby Appel (1987) put it, for Geoffroy Saint-Hilaire, "homologous parts were those parts in *different* animals which were 'essentially' the same, even though the parts might have different shapes and be employed for different purposes." Du Bois-Reymond's transition from physics to physiology soon did away with the sterile theorizing that *Naturphilosophie* represented for the aspiring rigorous scientist.

As Finkelstein notes, du Bois-Reymond's coming of age as a physiologist took place gradually during his years as Müller's student and trainee in Berlin, starting in 1839. He was first assigned to museum chores that would plague him for many years to come. In a letter to his friend Carl Ludwig dated 7 August 1849, du Bois-Reymond was forthright about his distaste for such duties: "Müller has kept me occupied at the museum, carrying out what is in his opinion the highest activity of the human intellect, namely, classifying fossil vermin" (Cranefield 1982). But in 1841 Müller, intuiting that the field of bioelectricity suited du Bois-Reymond's scholarly interests and research abilities, made amends by giving his trainee the recently published *Essai sur les phénomènes électriques des animaux* by Carlo Matteucci (1840). In it, Matteucci, who had made a name for himself in bioelectricity by seizing the torch from Galvani and Volta a generation earlier, gave a historical review of the topic and summarized his own research on frogs and torpedo fish. Indeed, du Bois-Reymond responded positively to his mentor's suggestion and embarked on a single-minded scientific pursuit that lasted his whole life.

By 1842 du Bois-Reymond had made landmark discoveries. He was among the first to see the importance of designing apparatus best suited to gain physiological insights. He discovered the law of the muscle current – how current flows over muscle fibres – and the negative current inflection of the action potential accompanying muscle contraction. The latter finding, Finkelstein stresses, was his most original. "The proof that electricity acts as a biological signal was one of the high points of modern physiology." Surpassing Matteucci's findings, which were based essentially on Galvani's approach of assessing bioelectrical effects indirectly through the all-or-none frog muscle twitches they produced, du Bois-Reymond aimed at measuring currents directly with skillfully handled electrodes and a sensitive galvanometer.

Although the frog remained du Bois-Reymond's preferred experimental model, he occasionally investigated other species as well. But he did so not out of curiosity about how diversely bioelectricity works in the animal world, but rather to reassure himself that his experiments were on the right track. His discovery of the law of muscle current, for instance, unnerved him so much that he felt compelled to test it in a wide variety of species: mammals, birds, reptiles, amphibians, fish, and crabs (du Bois-Reymond 1852). This was the extent of his comparative approach. As Otis (2007) remarked in her study of Müller and his students, "for the most part, they [among whom she included du Bois-Reymond and Helmholtz] rejected [Müller's] notion that comparing animals would reveal the way that life worked."

Although du Bois-Reymond could revel in his scientific achievements, he understood that the level of knowledge necessary to explain how the flux dynamics of bioelectricity might lead to human consciousness was beyond his reach and, he mused, would remain so even with future leaps of scientific progress. In a lecture delivered in Leipzig in 1872, he dwelt on the likely impossibility of solving the conundrum of the origin of life on earth and the emergence of consciousness from brain activity:

> But now there comes in, at some point in the development of life upon the earth which we cannot ascertain – the ascertainment of which does not concern us here – something new and extraordinary; something incomprehensible, again, as was the case with the essence of matter and force. The thread of intelligence, which stretches back into negatively-infinite time, is broken, and our natural science comes to a chasm across which is no bridge, over which no pinion can carry us: we are here at the other limit of our understanding. (Du Bois-Reymond 1874)

Du Bois-Reymond's materialistic vision of brain function was best expressed in his endorsement of zoologist Carl Vogt's physiological analogy, which had caused great distress to adherents of the Cartesian mind-body dualism:

> Take Carl Vogt's bold expression, which in 1850 introduced a sort of mental tournament: "All those capacities which we call mental activities are only functions of the brain; or, to use a rather homely expression, thought is to the brain what the bile is to the liver, or the urine to the

kidneys." The unscientific world [was] shocked at the simile, considering it to be an indignity to compare thought with the secretion of the kidneys. Physiology knows no such aesthetic discriminations of rank. In the view of physiology the kidney secretion is a scientific object of just the same dignity as the investigation of the eye, or the heart, or any so-called "nobler" organ.

For du Bois-Reymond it all came down to the fundamental elements of matter and force that are behind every physical phenomenon in the universe, including animal functions. They are knowable only to a degree: "But as regards the enigma what matter and force are, and how they are to be conceived, [the investigator] must resign himself once for all to the far more difficult confession – Ignorabimus!" Fortunately, some – the Einsteins of the next generations – were not prepared to submit to such a resignation.

Du Bois-Reymond's close friend Ernst Wilhelm von Brücke displayed more eclectic physiological interests than he did, but both shared a positivist attitude to the practice of physiology. Brücke completed his medicine at the University of Berlin in 1842 and remained Müller's assistant from 1845 to 1848. In 1849 he took a position at the University of Vienna, where he spent his entire professorial career. He had widely diverse interests: physiology of the eye, colour changes in chameleon and cephalopods, movements of the plant Mimosa, phonetics, philosophy, and aesthetics. He was particularly admired by the young Sigmund Freud, whom he mentored. As Freud's biographer noted, "For six years, between 1876 and 1882, [Freud] worked in his laboratory, solving the problems his revered professor set for him, to Brücke's evident satisfaction – and his own" (Gay 1998).

Helmholtz, on the other hand, shared many traits with his friend du Bois-Reymond. He dedicated himself to the experimental method and to the physico-chemical approach, and he also liked to reflect on his science in a philosophical or historical framework. He concurred with his friend that "the best way to explain a science is to tell its history" (quoted in Otis 2007), but he also humbly admitted the philosophical limitations of the experimental method. The friends both understood that "experiments do not generate facts; rather, they generate the contexts in which facts gain meaning" (Finkelstein 2013). Their careers differed in that Helmholtz, like Brücke, did not confine himself to a single research topic as du Bois-Reymond did, but embraced an eclectic set of physiological enquiries.

Helmholtz was born in Potsdam on the outskirts of Berlin into the family of a high school teacher and grew up in an intellectually stimulating household where books were aplenty and visitors of high stature. Among these were mathematicians and in particular the philosopher Immanuel Fichte, the son of the great philosopher Johann Gottlieb Fichte (Meulders 2010). Helmholtz heeded his father's advice to pay attention to philosophical ideas as guideposts to the study of nature. His dreams of a career as a mathematician and physicist were thwarted, however, by the family's financial circumstances, and he had to settle for medical studies at La Pépinière, the less costly military medical school in Berlin. One of his teachers was Johannes Müller, who detected Helmholtz's gifts and gradually attracted him to his laboratory (Meulders 2010). There he met du Bois-Reymond, who acclimated the newcomer to the lab rituals he so detested and guided him in his research apprenticeship.

Helmholtz lost no time in educating himself to the demands of rigorous experimentation. He first investigated the metabolic changes that take place when an isolated frog muscle contracts; he observed specific changes but failed to identify any chemicals involved. Next, using the same frog preparation, he asked how much heat was produced by contracting muscles and from what source. He succeeded in measuring the heat generated and also in showing that its source was within the muscle itself. And by ingenious means he measured the speed of conduction of impulses travelling in the motor nerve of a frog's gastrocnemius muscle. That these studies became nineteenth-century classics, his biographer Michel Meulders reminds us, hinges not only on the pioneering nature of Helmholtz's observations but also on the originality of his experimental methods: the isolation of the organ studied from the body so as to eliminate intrusive factors in the interpretation of data; the reliance on homologous control preparations to ensure the specificity of responses; and the repetition of experimental measurements to account for individual variations and to obtain statistical trends of the changes measured. These methods are taken for granted in physiological practice today, but were paradigm-changing at the time.

Meulders eloquently articulates the ethical standards that drove Helmholtz's scientific quest:

Helmholtz adhered sincerely to the three Greek pillars of experience of the "honest man": truth, goodness, and beauty. The search for truth was at the center of his scientific preoccupations. He did not have to bother himself too much with goodness because he was a man of duty and, as all true Prussians should be, highly disciplined. What caused him a problem was beauty because he did not know too well how to situate it in relation to science. Faithful to his philosophy of reducing if possible all observed phenomena to a single explanation, he gradually acquired up to the end of his career the conviction that the artist's perception was built up in the same way as the scientist's by unconscious sensory inferences. The intellectual processes were the same in both cases depending on memory and imagination by association. The difference between the artist and the scientist was that truth as seen by the artist was not the copy of a single object but the representation of an ideal.

As suggested, Helmholtz's greatest contribution lies in the field of sensory physiology. His treatise *Physiological Optics*, while serving as a repository of past achievements in vision research, brimmed with descriptions of his own discoveries. Not the least was his insightful understanding of colour vision. Thomas Young in England had proposed in 1801 that the mixing of three primary colours – red, green, blue – determined colour vision and that each primary colour worked in the eye through a specific receptor channel. (The way the eye processes colour, it turns out, differs from the way painters mix their primary colours – red, yellow, and blue.) Helmholtz went further, noting that "in physics one could only speak of the vibration frequency of light, whereas color is above all a psychological phenomenon" (Meulders 2010). As this quotation implies, Helmholtz can be considered a pioneer of "psychophysics" – the relationship between the physical properties of the stimulus and its psychological perception – before the word was coined by Gustav Theodor Fechner in 1860. Helmholtz explained how modulation of colour attributes are involved in the processing of colour vision in the eye and brain. Thus was born the Young-Helmholtz trichromatic theory of colour vision, which has largely stood the test of time to this day. Helmholtz also studied gaze and the ancillary role of eye movements. To achieve all this, he followed the lead of his friend du Bois-Reymond in constructing his own equipment

to best address the problems at hand – ophthalmometer, stereoscope, oph-
thalmotrope, and rotating disk, to name but a few.

The fourth adherent to the physiology manifesto, Carl Ludwig, stands out
from his colleagues in several ways. Born into a large family of modest means
near Göttingen, Ludwig struggled through his medical studies at the University of Marburg and, like du Bois-Reymond, had to submit to dull assignments as prosector of comparative anatomy before he could pursue a course
more in tune with his taste for physiology (Zimmer 1996). While in Marburg,
he undertook innovative studies of renal and circulatory functions that
earned him notice. In particular, he devised a new method of measuring
blood pressure. His rise among his peers earned him the chair of anatomy
and physiology at the University of Zürich in 1847. That same year he invented the kymograph, an apparatus destined to play an important role in
the practice of animal physiology. While Ludwig originally designed the
kymograph to measure blood pressure, its use spread to other functions,
particularly the recording of muscle contraction in real time. It was based
on the tracings of a pen tip on smoked paper wrapped around a rotating
cylinder. The movement of the pen reflects the activity of the functioning
tissue attached to it.

There then followed a ten-year interlude in Vienna during which Ludwig
made important discoveries in respiratory physiology. Thanks to dramatic
reforms spearheaded by the German government to promote excellence in
targetted scientific disciplines (Zimmer 1996), physiology was separated from
anatomy. As a result a physiological institute was created at the University
of Leipzig, and in 1865 Ludwig became its first chair of physiology.

In Leipzig, Ludwig distinguished himself from his three friends by attracting numerous students, who in turn spread the gospel of the new physiology
throughout Germany and elsewhere in the world. There were so many students in his lab that it was likened to a "factory of knowledge" (Fye 1986).
What made this possible was a host of historical circumstances that propelled
Germany forward as the most successful model of academic science and the
hub of physiological advances. Zimmer fills out the picture:

> [The Politicians] acknowledged that political power is dependent on
> industry, and that industry, in turn, is dependent on the development

Portrait of Carl Ludwig.
Wellcome Collection Catalogue Number M0011227.

of natural sciences. In view of this political, economic and social situation there were exceptionally good conditions for medical and physiological research during the entire professional lifetime of Carl Ludwig. As a result, German science and Physiology in particular flourished. The 19 independent German universities maintained by the kings or princes of the numerous small states constituting Germany guaranteed decentralization. Therefore, every state competed for the best scientists they could afford and attract. And these, in turn, could determine or negotiate their conditions in terms of laboratory space, equipment, personnel, and financial situation. Another factor was the structure of the German University system. Scientists in Germany became full-time researchers. After obtaining particular specialization, they could start an academic career and they were paid for their job by the respective state. This created a new scientific professionalism that developed its own standards of qualification and quality control. A particular qualification procedure was the "habilitation", the submission of a high-level scientific work based on original research. A "habilitation" was necessary to get the licence to lecture and to obtain academic promotion and appointment. Quality control was exerted and maintained by the requirement that the results of scientific research are published.

This new model, whereby young academics could develop their own scientific projects, build their own labs, receive adequate funding based on merit, and be assigned light to moderate teaching loads, was not followed up in the two other European powers: England and France. In English medical schools, as the historian Gerald Geison (1978) has masterfully shown, anatomy prevailed over physiology and the experimental method received very little support. The crusty, tradition-bound academic structure of British universities made matters worse. The only exception at the time was the appointment to a physiology professorship of Michael Foster at Cambridge in 1870. In the ensuing years Foster endeavoured to emulate Ludwig's Leipzig lab and attract talented students. Unfortunately, Foster's example had no entrainment effect, as the British government lacked the vision that led to Germany's reforms.

~

The French Academy presented similar systemic roadblocks. A young researcher, after his doctoral thesis, had to put up with a subservient role in the lab of the *maître*, and only the death of the *maître* opened up the possibility of running a lab, and not necessarily on the basis of merit. Claude Bernard (1813–1878), who, as mentioned, was the only French figure who measured up to the fabulous four Germans in terms of both experimental credo and accomplishments, had to endure this archaic, hierarchical system. Born to a family that owned a modest vineyard business in the Beaujolais region, Bernard moved to Lyon in 1832 to undertake a pharmacy apprentice-ship (Wise 2011). Disillusioned by the shaky scientific basis of the pharma-ceutical practices of his day, he drifted to the world of literature and wrote romantic plays in the manner of Victor Hugo. After negative criticism cut short his nascent literary career, Bernard moved to Paris in 1834 to begin medical studies. Around 1840, during his internship, he became research assistant to François Magendie.

Bernard's dream of dedicating his life to scientific research was thwarted in 1844 by academic setbacks and financial difficulties. Seeing the young man's potential, his mentor's team arranged for him to marry a woman from a wealthy family, in order to keep him in Paris and secure his career. He paid a price for the arranged marriage: his wife resented his complete devotion to his work and his resulting neglect of the family – and particularly opposed his practice of vivisection (Wise 2011). But his heavy investment in research enabled him to make one milestone discovery after another in experimental physiology. Ultimately, at the age of forty-two, he succeeded Magendie as chair of medicine in the Collège de France.

In 1865 Bernard published his landmark opus, *Introduction à l'étude de la médecine expérimentale*, in which he clearly and elegantly expounded his concept of the experimental method as it applied to medical physiology: "[It] is nothing more than a reasoning by which we methodically submit our ideas to the experience of facts." He postulated that the complexity of life phenomena dictated the development of techniques that met the demands of biological experimentation, as distinct from the techniques of physics and chemistry. As the book title suggests – and its contents confirm – Bernard did not separate physiology from medicine; it was as if physiology could not exist outside the medical field. This view was very different from that of the Germans, for whom physiology, although helpful to medicine, was an inde-pendent discipline. Also, in the chauvinistic and authoritarian fashion typical of his era in France, Bernard ignored his German colleagues' pioneering con-tributions to the theoretical foundations of experimental physiology, espe-cially those of du Bois-Reymond (1848) and Ludwig (1852), although the latter held more mechanistic views than Bernard.

Claude Bernard.
Wellcome Collection Catalogue Number M0010569.

Bernard's snobbish attitude contrasted starkly with the humble conditions under which he toiled. His success was the more remarkable for the limitations they imposed. As William Bynum (1994) explains:

[Bernard's career] was "extremely French in its trajectory. Unlike his German colleagues, Bernard never headed an institute, and, although he eventually was able to command adequate laboratory facilities, his early years were spent in sparsely equipped and sometimes even privately funded space. Nor was his research ever dependent on much sophisticated laboratory apparatus. Rather, his genius lay in his superb operative techniques, his capacity to keep experimental animals alive through the follow-up and interpretative parts of his investigations, and the elegant simplicity of his experimental designs. His major findings – the role of the liver in synthesizing glycogen and in keeping blood glucose levels within a defined range; the digestive functions of the pancreas; the vasodilator nerves; the site of action of poisons such as carbon monoxide and curare – were characteristically based on simple but compelling experimental evidence.

An important physiological principle, which Bernard articulated late in his career, is the constancy of the "milieu intérieur" (Bernard 1878). With this notion, Bernard provides an object lesson on how animals can function diversely. He makes the distinction between latent life (in dormancy), oscillating life (in cold-blooded animals susceptible to the external milieu), and constant life (in warm-blooded animals whose interior milieu is maintained independently of external conditions). In the latter case, Bernard adds that complex mechanisms must operate so that the interior milieu does not change around "living particles, fibres and cells."

This principle was taken up decades later by the American physiologist Walter B. Cannon (1871–1945), who enlarged it in his concept of homeostasis (Cannon 1932):

The coordinated physiological processes which maintain most of the steady states in the organism are so complex and so peculiar to living beings – involving, as they may, the brain and nerves, the heart, lungs, kidneys and spleen, all working cooperatively – that I have suggested a special designation [as opposed to *equilibria*] for these states, *home-ostasis*. The word does not imply something set and immobile, a stagnation. It means a condition – a condition which may vary, but is relatively constant.

Cannon studied at Harvard under Henry P. Bowditch (1840–1911), who had also spent time in Bernard's and Ludwig's labs after earning his medical degree. Although Bowditch enjoyed his stay in Paris, he found the French method very lacking in comparison to "the exhaustive way in which questions are treated by the German investigators" (Cannon 1922). In fact, Bowditch purchased physiology equipment in Germany to bring back to the Harvard Medical School, where he organized the first real physiological laboratory in the United States open to students. In this initiative, he joined the drive led by Michael Foster in England to modernize physiology along the German model in their respective countries.

If Bernard's lab could not boast the institutional support for medical physiology enjoyed by the Germans, it nonetheless produced a scientist who made significant contributions to animal physiology. Paul Bert (1833–1886) worked as Bernard's assistant before occupying the chair of comparative physiology at the Muséum d'histoire naturelle, where he embarked on his own eclectic research program. In addition to his foundational investigations on anaesthetic gases and high altitude physiology, Bert worked on various aspects of the physiology of the cuttlefish (Bert 1867) and on a broad survey of respiratory functions in the animal kingdom (Bert 1870). In the latter opus he outlined his philosophical approach to comparative physiology, stating that he wanted to venture further than his predecessors in two ways: "by applying experimentation to the explanation of facts of natural history, and ... by using facts observed in lower animals for the study of physiological or pathological problems faced by the human species." This was an early statement about the multifaceted roles that comparative physiology would play as it developed in the next century.

Although Bert dedicated his book on the comparative physiology of respiratory mechanisms to his old mentor, his approach departed from Bernard's in its comprehensive description of techniques and measuring instruments – including drawings of apparatus – as well as in its extensive bibliography evidencing unusual care in giving his predecessors their due.

Sketch portrait of Paul Bert.
From Popular Science Monthly, volume 33, July 1888.

He used a French adaptation of the kymograph to record a wide variety of respiratory processes from an impressive list of animals: sea cucumbers, bryozoans, sea squirts, boring clams, annelid worms, lobsters, insects, cuttlefish, lancelet (Amphioxus), hagfish, lamprey, fishes, tadpoles, caiman, turtles, snakes, ducks, dogs, humans. Bert's investigations uncovered mechanisms for pulmonary air inspiration and expiration and for gas exchange in the blood of aquatic as well as aerial animals, and led to many other discoveries. He showed how animal groups living in contrasting environments differ in their handling of respiratory challenges. Not surprisingly, only eleven years after the publication of Darwin's *Origin of Species* and its tepid reception by French biologists, Bert was silent on the possible evolutionary implications of his findings.

This silence of a physiologist such as Bert on Darwin's theory may be explained by the absence of any serious consideration of animal functions in *Origin of Species*. Darwin advanced natural selection as the cause of adaptation, but the word "adaptation" carried little functional weight; in Darwin's discourse, morphological traits were more apparent. The experimental biologist George Romanes (1848–1894), a close disciple and friend of Darwin from 1874 until the latter's death (Schwartz 1995), had somewhat strayed away from Darwin's thought by stating that "it is the office of natural selection to evolve adaptations – not therefore or necessarily to evolve species" (Romanes 1886). To evolve species, Romanes proposed the concept of physiological selection, whereby the variability of an organism's reproductive organs – either intrinsic or influenced by the environment – can determine its barrenness or fertility and therefore its extinction or fitness to propagate new variations. But Romanes was possessed of a narrow, if not strange definition of "physiological" that Bert and other animal physiologists could not identify with. At the time of Bert's investgations few if any physiologists were prepared to conceptualize "physiological adaptations" in an evolutionary mold, as animal physiology and evolutionary thought were both still struggling for recognition.

Bert's comparative investigations depended greatly on obtaining marine specimens at the marine laboratory of Arcachon near Bordeaux. Arcachon represented an early trend in the development of marine stations in France. As Jean-Louis Fischer (1980) explains: "It became evident that in order to study the marine fauna no collection could replace freshly captured material on shores at low tides or in open sea, and when physiological or embryological questions were tackled it seemed essential to travel on site to conduct such investigations." As an added bonus, investigators could watch the animals in their natural environment to give context for their experimental results. In this regard France was a pioneer in Europe, as the following timeline of openings of marine laboratories shows: Concarneau (1859), Arcachon

(1863), Roscoff (1872), Wimereux (1874), Luc-sur-Mer (1874), Sète (1879), Banyuls-sur-Mer (1880), Villefranche-sur-Mer (1882), Tatihou (1887), Le Portel (1888), Marseille-Endoume (1888), and Tamaris (1891) – not to mention, at France's doorstep, the Institut océanographique de Monaco (1906). Not all of them survived, but France's visionary understanding of the importance of marine stations for zoology was unequalled.

Despite the embarrassment of riches that French seaside labs presented, the promise of their support for animal physiology, as exemplified in Bert's practice, was not fulfilled. The investigators who assiduously visited the coastal stations favoured life history studies and evolutionary morphology. To make matters worse, the complex politics of station managers and personality conflicts kept marginalizing physiology. Arcachon and the short-lived Tamaris station were exceptions (Bange 2011). Although evolutionary morphology was the workhorse of the Naples Zoological Station, founded by the German Darwinian Anton Dohrn in 1872, that station was more welcoming than most to other zoological fields as well, and marine animal physiology was able to carve out a space there (see chapter 4). Despite the obstacles put in place by some staunch anti-physiologists, the opportunities to experiment on a great variety of marine animals at seaside labs widened the field of animal physiology and allowed it to thrive in the last two decades of the nineteenth century and into the next.

One important, if forgotten and controversial French "zoophysiologist" who made heavy use of seaside laboratories was Georges Bohn (1868–1948). He is representative not only of the kind of field work practised by the French for the sake of animal physiology, but also of the views held by French zoologists of that period in relation to comparative and evolutionary physiology. All that is known on the social background and formative years of Bohn is that he was one of the many students and disciples of Alfred Giard (1846–1908), an influential zoologist who founded the marine laboratory of Wimereux and who taught both Lamarckian transformism and Darwinian evolution at the Sorbonne. Bohn himself defended Lamarckism while holding his own chair at the Sorbonne.

Bohn's monograph on the respiratory system of crustaceans (Bohn 1901) was boldly subtitled *Essay on Evolutionary, Ethological and Phylogenetic Physiology*. It seems an extraordinary program for animal physiology at the dawn of the twentieth century, but one may ask why it separates evolutionary from phylogenetic pursuits. A Darwinian would not have taken this path. In his justification Bohn betrayed his Lamarckian bias: "It is imbued with the theory of evolution that I approach the physiological study of Crustaceans. I have sought to bring to light the influence of the external milieu, habitat, life habits (ethology) on the function and subsequently on the form, and thereby to follow the lineage of species (phylogeny); in brief I have attempted to do comparative, ethological and phylogenetic physiology."

If Bohn's Lamarckian approach is evident in the introduction to his monograph, it also signals the adoption of an environmental approach that was destined to pick up steam later in the century. Bohn also formally introduced the word *ethology*, which he understood as the study of animal behaviour in the natural environment and as an integral part of comparative physiology. But ethological practice, in his hands, predicated as it was on observations of the animal in his native habitat, precluded experimentation and exploration of specific functions, thus denying his version of ethology accessibility to physiological inquiry. A more lucid role for his ethology would have been to provide context to functioning animals.

This muddled perspective on the mission of comparative animal physiology pervades many texts of the time. It simply reflects on the debates attending the birth pains of the discipline. A prominent issue was that for "zoophysiologists" experimentation was less central than for medical physiologists. Bernard had criticized zoologists in this regard, somewhat arrogantly opposing his "active physiology" to their "passive zoology" (Debaz 2005). This attitude did not sit well with zoologists. Henri de Lacaze-Duthiers (1821–1901), a student of Milne-Edwards and founder of the marine laboratories in Roscoff and Banyuls-sur-Mer, for instance, rejected Bernard's criticism, arguing that the latter's definition of the experimental method was too narrow: "[E]xperimental action is deterministic, that is, the experimental process or operation exposes the well-known properties which allow the reproduction or termination of vital phenomena just as one reproduces or terminates chemical reactions" (Lacaze-Duthiers 1872). In the view of Lacaze-Duthiers and many of his followers: "[A]n experiment serves to test a theoretical induction or inference. The scientist induces a theory from the observation of phenomena, and experimentation lies in the verification of this induction. Therefore the experimental method consists in observing all phenomena and controlling the interpretations derived from them. This proceeds from our inability to control everything, to apprehend reality in a holistic manner" (Debaz 2005). The original proponent of this concept, the chemist Michel-Eugène Chevreuil (1786–1889), called it the *a priori* experimental method (Chevreuil 1866).

Science historian Karl Figlio (1977) goes to the heart of the matter when he states that Bernard placed "a total emphasis upon concepts and methodology, upon *modes of knowledge* rather than the *nature of beings*" (Figlio's italics). Figlio continues by exposing what he saw as the shortcomings of Bernard's conceptual framework:

> Knowledge is indistinguishable from the ability to place constraints upon an isolated field of study. Knowledge is control; we know because we master. The test of this knowledge is that the same limited succession of events will always follow the same imposition of conditions. Greater refinement of this technological ability will produce greater refinement of reproducibility, i.e., in entailing the same consequences. It also follows that the field of future investigation will be increasingly restricted, increasingly identified with the available technology, and increasingly separated from concern for the specialness of the object investigated.

One can appreciate here how this perceived lack of concern for the animal model created a conflict with animal physiologists for whom the particular physiological solutions adopted by a specific animal species facing its own set of environmental challenges must not be ignored. The heated debate frayed many egos in the zoologists' camp. As Josquin Debaz (2005) illustrates, historian of French science Harry Paul (1985) satirized it in the form of a male ego psychodrama, but the subtext is insightful:

> Since science was an exclusively male activity, insult was certainly added to injury in Bernard's categories, which might have conjured up out on the depths of the male ego the horror of a relegation of zoology to an area of female activity. If nature's secrets were to be penetrated, Coste, Daubrée, and Lacaze-Duthiers did not wish to be deprived of their epistemological phallus (experimentation). More important issues were at stake. Lacaze-Duthiers's assertion of the experimental rights of zoology, the best known of the defenses against Bernard's attack, is not much of

a surprise when one considers that he was a product and leading representative of the disciplinary matrix being challenged. Nor did it hurt his situation to defend the work of his powerful maître Henri Milne-Edwards, old-style zoologist, anatomist, and experimental physiologist, whose versatility probably irritated Bernard to some degree but not too deeply, for he could take comfort in Edwards's lack of originality.

Beyond this internecine quarrel loomed a larger issue, one that Bohn alluded to in his monograph: zoologists had not done enough to promote their brand of animal physiology and had left too much room for medical physiologists to intrude into their field. Comparative physiology suffered both from lack of consensus on its theoretical foundations and from a dearth of research results. To make his point on the shortfalls of comparative physiology, Bohn (1901) disconsolately exposed what happens when one leaves the field to medical physiologists:

> While there are countless works on human and higher vertebrate physiology, one book only, and unfinished at that, Krukenberg's *Vergleichend-Physiologische Vorträge* (1886), reports on publications in the field of invertebrate physiology. When one peruses the bibliographic indices found at the end of each chapter, one is struck by the fact that the few researches conducted on lower animals proceed generally from human physiologists. The latter often develop marvelous techniques, they arrive at absolutely rigorous results, but when they move out of their own field they remain imbued with anthropomorphic tendencies: like the ancient zoologists they reduce everything to humans. They explain simple phenomena occurring in a protozoan or a coelenterate through the prism of complex phenomena taking place in higher vertebrates.

It was twelve years after Bohn's lament before another book focused on invertebrate physiology appeared. The author of the opus, Hermann Jacques Jordan (1877–1943), was born in Paris to a German merchant and a Jewish mother. He studied under the great biologist Theodor Boveri in Würzburg and the eminent medical physiologist Eduard Pflüger in Bonn, who taught him the experimental method (Postma and Smit 1980). In 1898 he went to the Naples Zoological Station to do research under the supervision of the station's director, Anton Dohrn. Jordan's research there on the locomotion of the sea slug *Aplysia* earned him his doctorate back home in Bonn. As there was no academic employment for a comparative physiologist at the time, Jordan killed time as a lecturer in Zurich. In 1907 he finally took a professorship in Tübingen, where he completed his book, *Vergleichende Physiologie Wirbelloser Tiere* (Comparative Physiology of Invertebrate Animals, 1913).

Jordan's book was unfortunately the first volume – on feeding methods and digestive processes – of a treatise that never saw the light of day. In the year of the book's publication, Jordan moved to the University of Utrecht in the Netherlands, where he was offered a chair of comparative physiology, the first in that country. This move may have been motivated by his marriage to a Dutch woman originally from Rotterdam. In Utrecht he made pioneer contributions on the neuro-muscular system of crustaceans and attracted brilliant students, foremost among them the invertebrate physiologist Cornelis A.G. Wiersma (1905–1979) (of whom more in chapter 9). In May 1943, however, Jordan lost his academic position and went into hiding with his Jewish wife to elude the Nazis. He died of a massive stroke in his hiding place on 21 September 1943.

∽

Jordan's lab in Utrecht was financially supported by the Rockefeller Foundation, which was established in 1913. The same Rockefellers (father and son) created the Rockefeller Institute for Medical Research (later Rockefeller University) in New York City in 1901, an institution that counted among its faculty an important figure in this story, Jacques Loeb (1859–1924). Loeb was much admired by Georges Bohn, whose research program was largely inspired by Loeb's special kind of physiology, especially the study of behaviours associated with attraction to or repulsion from specific environmental stimuli (light, temperature, chemicals), otherwise known as tropisms.

Loeb's importance in our story stems from the central role he and his circle played in the conceptual turmoil that beset physiology at the turn of the twentieth century. Philip J. Pauly's 1987 biography of Loeb, on which much of the following discussion is based, provides a broad canvas of the biological ideas simmering in Loeb's era as well as a narrative of his life. Loeb was born Isaak Loeb to a German Jewish family in the Rhenish town of Mayen. After school graduation and as he was about to start his medical training at the University of Berlin, he changed his first name to signal his atheism and to reject the nationalistic "atmosphere of Bismarck's *Kulturkampf*" by adopting a French name (Pauly 1987). Loeb's exposure as a medical undergraduate to the brain function localization studies of Eduard Hitzig and Hermann Munk led to his embrace of physiology in 1881.

But in keeping with what is described as Loeb's contrarian and adversarial personality, in 1884 he chose a critic of the "brain localizers," the Strassburg physiologist Friedrich Goltz (1834–1902), to supervise his research thesis. From his frog experiments Goltz had concluded that no specific brain region had the monopoly of a cerebral function; instead, brain performance, he said, owed much to dynamic, adaptational processes capable to some extent of offsetting brain lesions. Loeb's thesis reinforced Goltz's point of view. As soon as he earned his medical degree in 1885, Loeb abandoned the medical system and, for want of a better position, experimented on dog brain lesions at the Berlin Agricultural College. Pauly traces Loeb's increasing alienation from the German physiology establishment in the years of his appointment as assistant at the Physiological Institute of the University of Würzburg (1886–88). This estrangement led to the proliferation of a string of unorthodox views over the years which pitted Loeb against many high-profile biologists and physiologists.

One profitable outcome of Loeb's stay in Würzburg was his intellectual exchange with the local botanist Julius von Sachs (1832–1897), one of the great pioneers of plant physiology. According to Pauly's account, Sachs influenced Loeb in two ways: first by inducing him to search in the animal world for the behavioural orientation to light (heliotropism) or gravity (geotropism) that the botanist had studied in plants; and second, by rallying Loeb to his view that plants and animals share fundamental, holistic physiological traits that justify their treatment in the framework of a *general physiology*. Loeb lost no time in studying tropisms in "lower animals" in order to validate this new physiological approach (Loeb 1890). Soon he was using the Naples Zoological Station as a research base to work on hydroids

Jacques Loeb in 1915.
National Library of Medicine Image B017356.

and related forms. It seemed that, to emulate Sachs's experimental approach, Loeb deliberately chose animals that eerily resembled marine plants. In the monograph that resulted from his stay in Naples (Loeb 1891), he introduced the term *heteromorphosis* to denote the induced disruption of the hydroid polarity during regeneration, which resulted in the production of two heads, at opposite ends of the animal.

Animal tropisms and heteromorphoses fed into Loeb's efforts toward control, and Pauly's biography stresses that aspect of his biological outlook. Loeb believed that, through experimental manipulation, he could control the behaviour or body organization of his animal models in predictable ways. This belief led him to an engineering approach to physiological inquiry, by which he meant that he was simulating the control that, say, a civil engineer exerts on his materials in the course of a building project. Far from him the notion – too romantic or metaphysical? – that animals are products of nature's engineering designs, sprung from the drawing board of some cosmic power or evolutionary process. No, Loeb substituted himself as the engineer, and as a biological engineer he sought to manipulate the ability of living matter to regulate itself in ways that would benefit mankind.

Loeb's engineering approach was markedly off course from that of the mainstream physiologists of the medical establishment, toward whom Loeb expressed resentment, if not scorn. Paul de Kruif, a noted microbiologist and popular science writer who knew Loeb at the Rockefeller Institute, quoted in his autobiography (1962) what the German felt about medical research: "'Medical science?' said Jacques Loeb, chuckling, 'Dat iss a contradiction in terms. Dere iss no such thing.' Likewise he had no patience with theoretical constructs in biology, insisting that 'Every philosopher is either a swindler or a fool.'"

Loeb's alienation from German physiology and his marriage to an American woman led to his emigration in 1891 to the United States, where he took academic or research posts successively at Bryn Mawr, Chicago, Berkeley, and the Rockefeller Institute for Medical Research in New York. On American soil he encountered a different physiological tradition than in his native country. In Germany biological disciplines were clearly divided between medical physiology and zoology/botany. Not yet so on American soil, where a new breed of biologists, spearheaded by Charles O. Whitman in Chicago, sought to include physiology as part of a set of "diverse specialties [that he] wanted to have working cooperatively together under the broader rubric of 'biology'" (Maienschein 1987).

A similar cooperative program had been attempted earlier at the newly created Johns Hopkins University. At Johns Hopkins, physiology landed in the biology program because the university's flagship – the medical school and affiliated hospital – had failed to materialize in time (Maienschein 1987). However, its recruited physiologist – Henry Newell Martin, a protégé of Michael Foster's physiological laboratory at Cambridge – stuck to the mammalian heart research of his alma mater and failed to integrate into Johns Hopkins's biological program. To make matters worse, the ripple effects of Martin's alcoholism cut short any hope of a future for animal physiology within a broader biological program, and the physiologists soon migrated to the burgeoning medical school.

While Whitman thought that physiology had better prospects in Chicago with Loeb, it became evident that much cacophony resonated in any discourse on what physiology represented to whom. Whitman's biological physiology – where physiology was just another level of explanation within the life sciences – was not echoed in the ever-shifting constructs that Loeb came up with to promote his brand of physiology. Indeed, one is dizzied by the succession of approaches that he advocated over his career in the United States – from engineering to mechanistic, holistic, and comparative stances. The contention grew to the point that: "In Whitman's mind Loeb was no longer a biologist, but merely a 'physiologist'; by the same token, biology was no longer a combination of all the life sciences, but the equivalent of 'experimental natural history'" (Pauly 1987).

By the mid-1890s Loeb's view of physiology centred around the anchoring concept of "living matter," a phrase meant to designate the fundamental properties of cells and tissues. The task of physiology, he claimed, was twofold. Comparative physiology was called to the task of apportioning what is common to all living matter and what represented special functions in subsets of organisms. The field of general physiology, on the other hand, was reserved for the pursuit of the basic physical-chemical constitution of cells.

Loeb was not isolated: others, especially among his German compatriots, had expounded comparable views. One of them, Max Verworn (1863–1921), may even be regarded as a direct competitor in the same race as Loeb. Verworn was born in Berlin, where he received all his education up to his doctoral thesis in 1887 (Cathcart 1922). He followed up with a medical degree at Jena in 1889, but shortly jettisoned medicine for zoology. He worked at the Villefranche station in the south of France and at the Naples Zoological Station, where he met Loeb. He held professorships in physiology in Jena, Göttingen, and Bonn.

In 1892 Verworn published a monograph in which he claimed that protoplasmic movements in amoeba and other protozoans share a mechanism with muscle fibres involving "contractile particles." The comparative approach he used allowed him to trace many functional phenomena of higher organisms in lower animals, even in single-celled animals. All that differed among them, he claimed, were the small details of functional implementation. This work adhered somewhat to Loeb's first task of physiology. In 1894 Verworn's *Allgemeine Physiologie* [*General Physiology*] was published, the first textbook in a field that he helped develop along with Loeb (Verworn 1899 for English translation). In it Verworn clearly stated that general physiology is synonymous with cell physiology, a statement that represented the second task of Loeb's physiology. All this, it seems, was accomplished before Loeb articulated such views in print. Even Loeb's forum for his views, the *Journal of General Physiology* which he founded in 1918, was largely preceded by Verworn's *Zeitschrift für Allgemeine Physiologie*, started in 1902.

Not surprisingly, from what we know of Loeb's abrasive personality, his response to Verworn's work was exceedingly hostile. Even though Verworn was said to be curiosity-driven and to exhibit a broad range of scientific and cultural interests that led to authoritative publications about the philosophy of science and art history (Cathcart 1922), Loeb considered him an "ignoramus" (Pauly 1987). Without indulging too deeply in psychological motives, one can surmise that Loeb's reaction was in part attributable to a defensive reflex for being beaten to the finish line. The lost race was made even more bitter by Loeb's conviction, not entirely unjustified, that Verworn's scientific work was inferior to his own.

If Loeb was antagonistic to Verworn, he nevertheless attracted several German-speaking biologists, who came to his labs in Berkeley and Pacific Grove in the Monterey Peninsula. This movement represented a reversal of the much more common flow of young Americans to German universities. One of these visitors around the year 1903 was the controversial physiologist and zoologist Theodor Beer (1866–1919). Beer was born in Vienna to a wealthy merchant and banker. After studying medicine in Vienna, Strasburg, and Heidelberg, he trained in ophthalmology and eventually worked at the Physiological Institute of Bern in Switzerland. The research he conducted at the Naples Zoological Station led to his landmark publications on how animal eyes actively accommodate their optics for long-distance (far-field) or short-distance vision (Beer 1894, 1898). He also made comparative studies on the organs of hearing in invertebrates (Beer 1899).

In 1896 Beer was appointed assistant professor of comparative physiology at the University of Vienna. Three years later, with fellow physiologists at the Naples Zoological Station, he published a "manifesto" calling for a harmonization of the terminology regarding sensory processes and reflexes based on the broader picture of the whole animal kingdom (Beer, Bethe, and Uexküll 1899). Thanks to its attempts to bring some "mechanistic" coherence to comparative (animal) psychology, this paper was influential in decades to come (Dzendolet 1967; Mildenberger 2006).

Shortly after Beer was promoted to associate professor in 1902, a scandal erupted that destroyed his career. He was accused of committing a sexual act with minor boys (Vyleta 2005). Just as a preliminary hearing of the case was scheduled, he decided to flee Austria. One of his ensuing peregrinations brought him to Berkeley, where he collaborated with Loeb. But, as Pauly (1987) relates, Beer, "who already apparently had deep psychological problems, abandoned laboratory work and settled in Pacific Grove as 'Count Hallenberg', refusing to acknowledge that he knew Loeb." When Beer returned to Vienna in 1905 to face trial, the fierce ambient antisemitism fuelled by Vienna's mayor, Karl Lueger, led to his conviction despite a lack of evidence (Vyleta 2005). His appeal having failed, his young pregnant wife committed suicide. Beer lived in exile in Switzerland, where he remarried, but he in turn committed suicide after a bad investment in Austro-Hungarian war bonds left him bankrupt.

If the promise for comparative animal physiology embodied in Beer's accomplishments was cut short, the field was not left without protagonists. In Germany, England, the United States, and other countries, a new generation of energetic researchers entered the field who introduced novel ways of approaching animal functions. They in turn consolidated the impact of the field by founding their own publishing outlets.

4

The Turning Point

~

How simple these questions seem to us now! Do bees perceive colors,
measure distances, orient to the sky and to the changing position of the sun?
Do they have an internal clock? Can a fish hear and differentiate tones? But each
one of us knows that the simplest is often the most difficult and that simplicity
is not only a mark of truth but also of genius.

~ Hansjochem Autrum (1982)

The second and third decades of the twentieth century were pivotal in constructing modern animal physiology and fostering the nascent discipline
through institutional channels and publishing outlets. The first salvo toward
these achievements was the publication of Hans Winterstein's *Handbuch der
vergleichende Physiologie* (1910–24). Never before had any animal physiologist
attempted to contain between two covers the expertise of so many on a dizzy-
ing variety of functional topics.

Winterstein (1879–1963) was born in Prague, where he studied medicine,
before moving on to Jena; but it was in Göttingen, under Max Verworn's supervision between 1903 and 1906, that he came into his own (Weber and
Loeschcke 1964). Like so many of his contemporaries, he paid visits to the
Naples Zoological Station to work on local marine animals. Between 1906 and
1927 he rose through the professorial ranks at the University of Rostock, and
then moved to Breslau University. Although Winterstein was a convert to the
Catholic religion, he was still a Jew in the mindset of the Nazis. As a result he

left Germany in 1933 and settled in Turkey, where he developed a physiological institute at the University of Istanbul. He spent his last years in Munich.

From early on in his research career, Winterstein focused on the physiological regulation of breathing. At first he examined how squids and other invertebrates, in addition to fishes and frogs, control their aquatic breathing, but he soon turned to a more fundamental approach in mammals and engaged increasingly in medical physiology. It was in this context that he developed his famous "reaction theory," according to which breathing is stimulated in a reflex response to low-oxygen ambient conditions through a chemoreceptor mechanism (Winterstein 1911).

How he came to edit his *Handbook of Comparative Physiology* is a mystery, as he wrote no introduction in which the genesis of the project could have been explained. He left no memoir mentioning it and no commentator on his life alluded to it. Winterstein was thirty-one and had just been promoted to full professorship in Rostock when the book project took form. If Winterstein can be said to have given birth to the *Handbook*, it is likely that Max Verworn acted as the midwife. Weber and Loeschcke (1964) emphasized Verworn's enormous influence over Winterstein, especially with regard to the comparative approach to animal physiology. Winterstein himself stated that Verworn's *General Physiology* was his bible. When Winterstein moved to Rostock, he took along the philosophy and practice of Verworn's lab and implanted them in his own fledgling physiological research. So Verworn could have acted as source of inspiration as well as guiding hand in the preparation of the *Handbook*.

But no matter how carefully one plans a multi-author book, the editor can lose control of the implementation, and the *Handbook* is a good example. It takes on a sprawling dimension – eight volumes – and an editor has to put up with the whims and egos of contributing authors. As a result Winterstein had to shuffle the contents of the volumes to ensure their timely appearance. A reviewer in the journal *Science* (vol. 40, issue 1018, p. 28, 1914) complained that the text was being issued "in fragments, prepared successively or simultaneously by different authors on quite unrelated topics." While putting the volumes together seemed rather chaotic, however, the treatment of the various topics, in typical German tradition, was systematic and exhaustive. The *Handbook* was considered a watershed for the field, and a reviewer in the British journal

Nature (volume 84, p. 102, 1910) noted that with its publication "the growing science of comparative physiology [was] receiving its due share of attention."

But if the *Handbook* enclosed the sum of knowledge about animal functions throughout the animal kingdom, it missed the opportunity to state the mission and specific goals of comparative physiology. This lack of intellectual depth on the part of Winterstein was soon to be made up for by the new focus and impetus of a fresh cohort of practitioners. Among the factors that led to the reprogramming of animal physiology were a growing awareness on the part of founding societies and publishing outlets dedicated to the discipline's mission of the concepts of evolutionary physiology, organism-centred biology, and environmental fitness; and a growing call for the field to ensure its independence from medical physiology. In the following pages we will analyse these factors.

One brooding question that occasionally surfaced in the previous chapter was how the comparison of animal functions can lead to evolutionary insights. The year before the launch of Winterstein's *Handbook*, two articles by the British physiologist Keith Lucas (1879–1916) decried the lack of expressed interest on the part of physiologists for Darwinian evolution (Lucas 1909a, b). Lucas, a young Cambridge-educated physiologist with a knack for technical wizardry who died during the World War I in the midair collision of his plane (Horace Darwin 1919), uttered this lamentable fact: "There is no break in the history of physiology to mark pre-Darwinian from the post-Darwinian period. Questions of function have never been called in to help in tracing the course of evolution, and the idea of evolution has given no aid in interpretation of the known facts of function. If the hypothesis of evolution were tomorrow to be proved untenable, physiologists would scarcely be concerned."

It is true that physiology's estrangement from evolutionary ideas could be blamed partly on the traditional collusion of physiology with medicine. But "the preoccupation of physiology with the study of man," in Lucas's words, not only distracted physiologists from embracing the comparative approach but also made them miss out on the homological classification of animals that became the exclusive purview of morphologists. "If physiologists had felt that the comparative study of function could form a science really essential to the understanding of evolution," Lucas argued, "it is hard to believe that they would not have hastened to remove the reproach so commonly made against them, that their animal kingdom comprises only the frog, the rabbit, the cat and the dog" (Lucas 1909a). As things stood, comparative physiology arrived too late in the field, and comparative anatomy had scooped the territory.

Lucas made a distinction between functional capability and the normal behaviour of cells or tissues. Functional capability should be the evolutionary marker, but Lucas complained about the paucity of data so "that the investigation of functional capability should be begun in a conscious and systematic manner." Now, how to go about studying phylogeny of function? "The only practicable method for the study of any function will be to investigate that function first in some cell in which it appears in a highly elaborated state, and with the help of experience and technique so gained to trace it back through succeeding degrees of less specialisation and greater obscurity" (Lucas 1909b). As this and subsequent chapters will illustrate, succeeding generations were attentive to this advice.

Organism-centred biology as a concept, Jan Baedke (2018) tells us, flourished in the 1910s and 1920s thanks to essays produced by numerous biologists and philosophers of science mainly from Germany and Great Britain. Baedke encapsulates the concept thus:

> This idea rests on the argument that many (if not all) biological processes cannot be investigated effectively without considering the causally efficacious unit of the organism, which not only transcends the properties of its interacting parts but mediates its material organization in coordination with environmental cues, constructs its material and social environment, and assembles with other organisms to form new kinds of individuals, among other things.

The concept, as this quotation makes clear, is multifaceted, but we will restrict this discussion to the aspects most relevant to animal physiology. But for a start, we might ask what was meant by the "individual" animal? The renowned British biologist Julian Huxley gave a fairly representative description in his 1912 work *The Individual in the Animal Kingdom*:

> First comes the minimum conception of an individual; the individual must have heterogeneous parts, whose function only gains full signifi-

cance when considered in relation to the whole; it must have some independence of the forces of inorganic nature; and it must work, and work after such a fashion that it, or a new individual formed from part of its substance, continues able to work in a similar way.

This description speaks of a functionalist outlook according to which the whole is greater than the sum of its parts by virtue of the complexity of an animal's organization. It also implies that the animal does not lose its identity in its external environment while interacting with it. Huxley states finally that the individual's offspring – through budding or sexual reproduction – displays the same "personality traits" and modes of activity as the progenitor individual. And if, as happens in certain species, the buds do not separate from the progenitor individual and take on specialized functions, one may end up with a colony that acts as a superorganism whose special identity depends on the coordination of the constituent buds, otherwise known as zooids.

In the 1910s there was a new awareness of the importance and extent of the environment in interpreting physiological phenomena. It did not translate into the "environmental physiology" that is practised today, but the discussions then taking place induced physiologists to add the physico-chemical properties of the external environment to the toolbox of explanatory factors that could account for physiological responses.

One physiologist/biochemist who stirred up such discussions was Lawrence J. Henderson (1878–1942). Born in Lynn, Massachusetts, Henderson received all his higher education at Harvard, and after postdoctoral studies in the laboratory of protein biochemist Franz Hofmeister (1850–1922) in Strasburg, he commenced his career at Harvard Medical School (Cannon 1943). His pioneering studies of acid-base regulation in organisms and the role of carbonic and phosphoric acids in this regard prepared him for a reflection on the role of the inorganic environment in living organisms. The historian and philosopher of science Iris Fry (1996) has remarked that "Henderson's research in biochemistry was instrumental in the establishment of the concept of the living organism as a self-regulating system that maintains dynamic equilibria." By the time he published his book *The Fitness of the Environment* in 1913, Henderson was an assistant professor of biological chemistry – later re-baptized as biochemistry – and only thirty-five years old.

Henderson argued that the physico-chemical environment of living organisms was a good fit for the emergence and sustenance of life. In fact, according to Fry (1996), "Henderson came to see the *specific constitution of the environment* as of crucial significance for life." The special properties of carbonic acid and water, for example, facilitate the equilibrium of an animal's internal fluids near acid-base neutrality (around pH 7.3). No wonder, then, that the internal fluids of lower animals reflect largely the salt composition of sea water, itself regulated at a slightly more alkaline pH, and that the blood of more complex animals follows the neutrality rule even if its salt composition does not adhere as stringently to that of sea water. In the Hendersonian concept, the idea that life originated in the sea is no-brainer.

"Henderson's conception of the environment," Fry (1996) emphasizes, "was not limited to a specific, local environment to which each organism has to adapt." Indeed he dealt with the higher level of the ecosystem of planet Earth and even with the physico-chemical features of the cosmos that led to the make-up of our planet. In stark contrast, the Scottish physiologist John Scott Haldane (1860–1936) narrowed his approach when he used the complex physiology of breathing as an object lesson in showcasing the relationship between organism and environment.

Born and educated in Edinburgh, Haldane eventually became a Fellow of New College, Oxford. A specialist in respiratory physiology, he experimented on himself and members of his family – which included his famous son the geneticist J.B.S. Haldane and his novelist daughter Naomi Mitchison – out of hesitation to use conventional animal models. Among his numerous discoveries and inventions are the earliest gas mask design and the Haldane effect – the release of carbon dioxide from the blood's haemoglobin when the latter binds oxygen (oxygenation). His experience as an observer of respiratory functions led him to expound his philosophy in his book *Organism and Environment as Illustrated by the Physiology of Breathing* (1917).

Here Haldane argued that neither vitalistic nor mechanistic theories can account for the relations between organisms and their environment. To counter the mechanistic theory – which, as we have noted in earlier chapters, had held sway for over a hundred years – Haldane had this to say:

A living organism has, in truth, but little resemblance to an ordinary machine. The individual parts of the latter are stable, within very wide limits of immediate environment, and in no way dependent on whether the machine is in action or at rest. This stability does not exist in the living organism. We find, it is true, that the living organism may react in a constant manner to a given change, just as a machine might do; but on investigation this turns out to be because the internal environment is at the time constant or "normal."

An important point made by Haldane concerns the distinction in the mind of physiologists between the impact of the external and the internal (blood, for example) milieu on the functioning of organisms. Haldane refused to draw such a distinction:

We cannot draw any complete line of separation between the regulation of the internal and that of the external environment; for evidently the one is complementary to, and indispensable to, the other. Regulation of the external environment is in fact only the outward extension of regulation of the internal environment, and the ultimate dependence on the external environment of the organs which regulate it is as evident as their more immediate dependence on the internal environment.

Haldane roots the research mission of animal physiology in the recognition of organisms as individuals: "The ground hypothesis or conception is that each detail of organic structure, composition, and activity is a manifestation or expression of the life of the organism regarded as a separate and persistent whole. We have therefore to make use of this hypothesis as a tool for investigation, just as the physicist uses the conceptions of mass and energy, or the chemist the atomic theory." From this credo he traces what he thinks should be the future of physiology:

The bane of physiology in the past has been inexact measurement and imperfect observation. The new physiology will be different. Its measurements and observations will be more exact, and, as has been shown in the previous lectures from actual instances, of a delicacy often far exceeding that of existing physical and chemical methods. But the observations and measurements will not be of phenomena which if isolated are mere illusions. The new physiology will not be content with causes, but will seek out the organisation of which "causes" are only the outward appearance.

In short, Haldane here makes an appeal for a physiology based on systems. Breathing, in his example, cannot be comprehended without taking into account how the components – air-inhaling and -exhaling apparatus, gas transfer to and transport in blood, oxygen dumping in tissues, and so forth – are functionally integrated into a whole, the breathing system. This was a program for the future of physiology in general, but how was it to be implemented in comparative animal physiology?

Haldane represented many British scientists who tended to be highly individualistic, if not outright eccentric. Their sense of belonging to a community of scientists of similar ilk was rather loose. Even when associated with an academic institution, they showed a predilection for setting up a lab in their own homes. Perhaps because their discipline had gone through several birth pains, as related in previous chapters, animal physiologists, out of insecurity, felt a greater need to work as a closely knit community. This meant, of course, founding scientific societies to promote their common goals. Given the peculiar logistics of their trade, however, what with the necessity to set up a lab where the animals of interest reside, especially marine species, it also meant that animal physiologists ended up rubbing elbows on lab benches of marine field stations.

∾

We have frequently alluded in previous chapters to zoologists and animal physiologists visiting French marine stations or Naples. Now it is time to examine the coalescence of this trend in the years leading up to the 1920s. The two institutions that exerted a major impact in this regard are the Naples Zoological Station and the Plymouth Laboratory. The Naples station was founded in 1872 by German zoologist Anton Dohrn (1840–1909) to promote Darwinism through zoological and morphological research on local marine animals. A new building to house comparative animal physiologists and "physiological chemists" (the proto-biochemists) opened its doors in 1906 (Ghiretti 1985).

Francesco Ghiretti (1916–2002), himself a physiological chemist who worked at the Naples station, recalled the Italian pioneering physiologists

Filippo Bottazzi.
From the archives of the Wellcome Institute.

who made important contributions there. He drew special attention to Filippo Bottazzi (1867–1941) and Silvestro Baglioni (1876–1957). Bottazzi, who was director of the station from 1915 to 1923, introduced the freezing point depression technique to measure osmotic concentrations of animal fluids (Bottazzi 1897), and the notion of osmoregulation *versus* osmoconformism to denote aquatic animals that keep their internal fluid osmotic concentration constant or fluctuating with the external milieu, respectively (Bottazzi 1908). Unfortunately, Bottazzi's strong affiliation with the fascists has done disservice to his reputation as a scientist (Stanzione 2011). Baglioni is best known for his work on sensory physiology and his massive monograph on the comparative physiology of the nervous system in Winterstein's *Handbuch der vergleichenden Physiologie* (Baglioni 1913).

As a measure of the impact of the Neapolitan *stazione* on comparative physiology, almost all the thirty-one contributors to Winterstein's *Handbuch* had worked at the station for some time. It was as if the marine fauna of the Bay of Naples accounted for what was hot in comparative physiological research at the time. The rich diversity of animals amenable to fruitful research was indeed extraordinary. One animal group, however, stood out from the rest and was adopted by many as their pet animal model: the cephalopods (squids, octopus, and cuttlefish). Even when researchers such as Bottazzi and Baglioni focused on fish, crustaceans, and sea slugs, they could not help also trying their hand at cephalopods.

From a compilation by Ariane Dröscher (*Octopus research at the Stazione Zoologica [1873–ca 1964]*), I estimate that no fewer than 175 publications related to cephalopod physiology and conducted at the station were published between 1890 and 1960. Several important physiologists occupied lab benches there – tables as they were called in Naples – from the 1910s to the 1930s, not least the Nobel Prize winner Otto Warburg (1883–1970). Using the sea urchin, Warburg "made the classic discovery that upon fertilization the rate of respiration of eggs rises as much as six-fold" (Ghiretti 1985). But certainly the researcher who made the Naples station especially famous for cephalopod physiology was John Zachary Young (1907–1997). J.Z., as he was called, could trace his ancestry to the famous British scientist Thomas Young (1773–1829) on his father's side and to the Lloyds banking family on his mother's (Boycott 1998). Educated at Magdalen College in Oxford, J.Z. was chosen to occupy the Oxford table at the Naples station in 1928–29, when he was

merely twenty-one. The course of his scientific career was steered in Naples
thanks to an Italian with whom he collaborated, Enrico Sereni (1900–1931).

Born in Rome to an intellectual Jewish family, Sereni spoke French, German, and English fluently by the age of twelve (De Leo 2008). His precocious
enrolment at the Medical School of the University of Rome was interrupted
in 1917 when he joined the military and fought in the last months of World
War I. After participating in medical teams struggling to stem the Spanish
flu epidemics, Sereni conducted research under several supervisors, including Baglioni (mentioned earlier) and at University College London, with the
renowned muscle physiologist A.V. Hill. In keeping with his precociousness,
at the age of twenty-five Sereni was appointed head of the physiological laboratory of the Naples Zoological Station. There he started a series of experimental studies on the physiological control of the remarkably fast skin
colour changes in cephalopods and on their nervous system function. Angela
de Leo (2008) explains why cephalopods were such popular animal models:

> Among invertebrates, cephalopods were the most suitable for labora-
> tory research for many practical and functional reasons. Firstly, they
> could be easily found in the Mediterranean, especially in the Gulf of
> Naples … and consequently they could be bought at a low cost. More-
> over, because of their ability to survive, also in hard conditions, they
> were particularly suitable for experiments with poisons or involving
> the isolation of parts of the nervous system. Thirdly, from the functional
> viewpoint, they presented some peculiarities that distinguished them
> from the rest of invertebrates[,] making them closer in some ways to
> vertebrate animals.

Learning how cephalopods work thus became a favourite pastime at the
station. What is salient in these animals is the way they use their fast changes
of skin colour and asperity to effect stunning camouflage or to communicate
with each other, all activities controlled by their impressive brain. It is as if
they wear their emotions and intelligence on their skin! But when J.Z. arrived
in Naples in 1928, cephalopods were not on his radar. He had planned to investigate the autonomic nervous system of fishes. Somehow Sereni attracted
him to his circle and their collaboration led to studies mapping the cephalopod peripheral nervous system by the selective section of nerve fibres and

John Zachary Young. Wellcome
Collection Catalogue Number L0012873.

degeneration/regeneration events (De Leo 2008). In the process Young discovered the giant axon of cephalopods, which became the cellular model for
understanding how the excitation signal is propagated along axons and how
sodium and potassium ions across the axon membrane determine the membrane's electrical potential (Keynes 2005).

More than a year after Young returned to Oxford, on 11 March 1931, Sereni's
wife found her husband dead in his bathtub (Boycott 1998; De Leo 2008).
The circumstances of Sereni's premature death were never made clear, and
a conspiracy of silence on the part of authorities or the family was suspected.
Suicide was touted, but so was murder, as Sereni counted among the leaders

of the antifascist movement in Naples. The "halo of mystery" (De Leo 2008) was never dissipated. While the death cut short Sereni's soaring career, Young built his own career on the cephalopod nervous system over many decades. This turn in his research field was not the only influence of Naples. "J.Z.'s own assessment of his undergraduate career, given in old age," Brian Boycott remarked, "was that he must have been very dull. If that were true, a year in Naples worked a remarkable change. Back in Oxford he sparkled." Thus did Naples work its magic, it seems, for many other visitors to the *stazione*.

J.Z. spent time also at a biological station closer to home. If not on a scale comparable to Naples, the Plymouth Laboratory in England did its share to rouse British zoology from somnolence and set it on a vibrant program of experimental zoology. Historian of biology Steindór Erlingsson (2009), in relating the role played by the Plymouth Laboratory in the development of experimental zoology in Great Britain, reminds us of what the field encompassed at the turn of the century. In Europe as well as in the United States it was then acknowledged to include experimental embryology, comparative physiology, and general physiology. Experimental embryology, which had a high profile in those days, aimed at analysing the mechanisms subtending the development and growth of organisms. General physiology, as we saw in the previous chapter, aimed at identifying the basic, common functional activities of "living matter." But comparative physiology – by virtue of its focus on animal diversity and its evolutionary and environmental dimensions – could best claim to help us understand how animals work.

Erlingsson's archival documentation clearly shows how the Plymouth Laboratory, founded in 1888 as a research outpost of the Marine Biological Association of the United Kingdom (MBA), was not able to create its department of physiology until 1920. British zoologists continued to enjoy the convenience of the facilities at the Naples Zoological Station and felt no urge to develop a similar program of experimental zoology on their soil. But World War I wrought havoc on the Neapolitan landmark, and its postwar financial difficulties made the site less attractive to British zoologists. In addition, financial institutions in the United Kingdom, governmental or otherwise, found it more expedient to bolster Plymouth than Naples, whose station, after all, was run by Germans, the recent enemy toward whom resentment still ran high.

The archives unearthed by Erlingsson (2009) contain a letter by British zoologist Stanley Gardiner expressing his hopes for "one big Marine Laboratory where zoologists, physiologists, biochemists, and oceanographers can all work at their problems together," adding "we look to Plymouth to fill this need." The biochemist and Nobel Prize winner Frederick G. Hopkins (1861–1947) narrowed Plymouth's mission down when he emphasized the need to work on lower animals whose functions are open books compared with the physiological systems of mammals, which at the time were less amenable to satisfactory analysis (Erlingsson, 2009). The idea, in Erlingsson's words, was that "studying the properties of marine organisms would enable wider generalizations in physiology and would provide the simplest way to settle many of the discipline's problems." What this mandate, pushed by medical physiologists not comparative ones, spoke of was an attitude that emphasized commonalities of function rather than diversity of physiological solutions to specific challenges met in individual species' local environments.

It is telling of "the shortage of non-medical animal physiologists in Britain at the time" (Erlingsson 2009) that the first director of the physiology department at the Plymouth Laboratory was a plant physiologist. In this regard the United Kingdom lagged far behind continental Europe. The problem extended to the hiring of staff. In the early years of the physiology department, the scientific output for animal physiology depended largely on visitors from outside the Plymouth campus. Two comparative physiologists in the early twenties contributed in particular to the Plymouth station's scientific output: a staff member, Carl F.A. Pantin (1899–1967), and a visitor, Lancelot Hogben (1895–1975).

Details of Pantin's biography were recorded by his colleague Frederick Russell (1968), and an account of his scientific research after 1935 has been provided by Michel Anctil (2015). Here we touch only on Pantin's early career. Born to a businessman father and a mother boasting a distinguished German musician among her ancestors (Carl Friedrich Abel), Pantin was attracted to zoology from his early teens. While serving as officer cadet of the Royal Engineers in the last months of World War I, he developed an interest in physics, and as a result he planned to study physics at Cambridge in 1919. But almost instantly he switched to physiology and zoology. Physics served him well in his budding research career, however, in that his knowledge of the subject allowed him to design clever experimental setups based on sound physical and mathematical principles.

Pantin joined the Plymouth Laboratory's physiology department in 1922 and left it in 1929, when he obtained a fellowship at Trinity College Cambridge. He immediately embarked on an ambitious investigation of external and internal factors affecting the contractility of the marine unicellular *Amoeba* (amoeboid movement), which resulted in a series of influential papers between 1923 and 1931 – *On the Physiology of Amoeboid Movement*. This and other research at the Plymouth Laboratory brought about two insightful essays of relevance to the advancement of comparative animal physiology.

In the first essay, Pantin (1931) reflected on animal body fluids, as their composition – almost akin to pockets of sea water in many marine invertebrates – evolved in others as a result of environmental osmotic challenges such as freshwater and aerial. He was among the first to stress the importance of the semi-permeable properties of biological membranes for shaping the ion composition and osmotic steady state in the fluids circulating around cells and in those within the cells. In so doing, Pantin also hinted at another property of membranes: active uptake of ions, later destined to gain status in our understanding of physiological regulation.

In the second essay Pantin (1932) dealt with what later became an iconic topic for animal physiologists: physiological adaptation. While adaptation is a generator of functional diversity, Pantin recognized that "the organism can never be infinitely plastic. All its structures are makeshifts which meet environmental requirements within the limits of the standard parts available for their construction." Pantin sounded a cautionary note about the determination of genuine adaptations. After all, any assertion of the existence of an adaptation is generally based on circumstantial evidence, and one needs all the hard evidence one can summon to bolster one's claim. Blood chemistry was a case in point: "Thus if we say that the ionic composition of the blood of an animal is adapted to the maintenance of the tissue-cells, it is, in fact, the surface membranes and excretory organs that have actually undergone evolutionary adaptation. Consequently adaptive significance of blood composition cannot be discussed till we have adequate knowledge of the structures which maintain it."

Pantin set out the conditions under which sound determination of physiological adaptations could best be achieved:

In all such cases what we require is more accurate description of the organism and its environment. Only in the field can the conditions of existence of the animal be truly determined. But to determine the adaptational significance of its characters to these conditions, their physical nature must he analysed in the laboratory, even though the experiments may seem far removed from the actual conditions of the animal.

～

Pantin's reproach to zoology, whereby circumstantial evidence too closely resembled speculation at the expense of experimentation, had a supporter in a colleague who actually sat with him at the Plymouth Laboratory: Lancelot Hogben. But whereas Pantin possessed wealth, charm, and warmth of feeling, and played the academic game deftly (Russell 1968), Hogben came from a poor background, inherited an unbending attitude on moral principles from his clergyman father, and was sharp in his criticism of colleagues and friends alike (Wells 1978). But equally, as George P. Wells observed, "Hogben was a brilliant biologist, a stimulating and indefatigable teacher, and a famous writer." This man not only made substantial contributions to comparative physiology early in his career but was also a pillar in the formation of the *Society of Experimental Biology* and the *British Journal of Experimental Biology* – both institutions still thriving today.

Born in Portsmouth to the family of a fundamentalist clergyman, Hogben grew to jettison religion and replace it with a deeply felt social conscience (his private ethics) and a career as a rational, mechanistic biologist (his public persona). He had just graduated from Trinity College Cambridge, when in 1917 he was imprisoned for several months for his stand as a conscientious objector with regard to military service. In 1919 he was appointed assistant lecturer in zoology at London University and in 1920 lecturer at the Imperial College of Science (Erlingsson 2016). During these years Hogben launched his research career working on cytological problems, but he soon grew disenchanted with this field, finding it slow at rallying experimentation.

Erlingsson (2016), who gained access to precious archives, charts the path that led to Hogben's embrace of experimental biology, especially comparative physiology. Hired in 1922 as a cytologist at the University of Edinburgh,

Hogben immediately switched to a research program on physiological control of pigmentary colour change in amphibians. This was no whim; already in 1919–20 he had collaborated with his newfound friend Julian Huxley on the metamorphosis-inducing effect of thyroid extract in the axolotl (Huxley 1920). In Edinburgh he followed up this line of enquiry (hormonal control) by examining the effect of surgically removing the hypophysis (pituitary gland) and of injecting pituitary extracts on the frog skin colour response (Hogben 1924). This response is known to help frogs blend chromatically with their background to elude the gaze of their predators.

Hogben stood at the cusp of a new subfield of comparative physiology. Endocrinology – the study of internal secretion of hormones and their actions – had just emerged as a subfield of medical physiology at the dawn of the twentieth century. Hogben's work on the pituitary counted among the pioneering contributions in a field that as yet had no formal status. Even though comprehensive reviews of hormonal effects on invertebrates were soon to appear (see chapter 10), comparative endocrinology rose to disciplinary status only on the occasion of the first symposium on the subject held in 1954 in Liverpool (Gorbman 1993).

Hogben ended up at the Plymouth Laboratory as the result of his uneasiness with the clinically oriented physiology department in Edinburgh and his desire to work in a more zoology-friendly environment (Erlingsson 2016). While still based in Edinburgh, during his spring and summer vacations between 1923 and 1925, he spent time in Plymouth, where he focused on invertebrate physiology, particularly on muscle activity and blood oxygen carriers such as haemoglobin and haemocyanin. More important, his interactions with physiologically inclined colleagues and students in Plymouth led to his writing a textbook of comparative physiology (Hogben 1926). As Hogben noted in the preface: "There is, so far as I know, no work in English which aims at giving an account of the physiology of the lower organisms." With his usual inexhaustible energy he went on to fill the void.

Hogben saw his *Comparative Physiology* as covering the new advances in the field where Winterstein's *Handbuch* had left off in 1912 as far as its literature survey was concerned. He summarized the breadth of his coverage thus: "[W]e shall consider first the characteristic activities which living organisms display; second, the sources of energy which lie behind these activities; third, the way in which the activities of an organism are co-ordinated with the changing conditions of the external world; and finally, the means by which a new animate unit is brought into being." For a thirty-one-year-old scholar, he had accomplished an incredible amount.

But already in 1923, when twenty-eight, Hogben was in the first tier of a momentous event for his scientific field: the creation of the *Journal of Experimental Biology* (JEB). G.P. Wells (1976), the zoologist son of the famous writer H.G. Wells, wrote that at the time "the younger zoologists ... most urgently needed a Society, and a *Journal*, to cater for the experimental aspects of their subject." And he added with wry humour: "The reason is that a great darkness had settled on the majority of British zoologists in the early years of this [the twentieth] century. They became obsessed by comparative anatomy and descriptive embryology, and by the possible evolutionary relationships of the animals whose corpses they studied." British zoology needed to catch up with Germany and America where, as discussed earlier, experimental biology had already been embraced.

But if Germany and the United States had steamed ahead, they had not created a banner around which zoologically minded physiologists could rally. The *Journal of Experimental Zoology*, founded in 1904 by the famous American fruit fly geneticist Thomas H. Morgan and cytologist Edmund B. Wilson – both at Columbia University – had raised expectations. As it turned out, the editorial board of the American journal was loaded with cytologists and experimental embryologists, and comparative physiology, an afterthought, never took off until later in the century. Neither did the foundation in 1928 of the journal *Physiological Zoology* by University of Chicago developmental biologist Charles Manning Child (1869–1954) at first help the cause of comparative physiology, for similar reasons.

It was incumbent on the new generation of British biologists to show the way. Erlingsson (2013) has dealt in detail with the way the Journal and the Society of Experimental Biology came into being. The journal appeared first, and the society was founded in short order to support the journal and to "further the cause of the experimental approach." According to G.P. Wells (1976), the idea of the journal had been brewing in Hogben's mind for some time. During his tenure in Edinburgh he became surrounded by like-minded colleagues, among whom his boss Frank Crew, Julian Huxley, and J.B.S. Haldane – the son of physiologist J.S. Haldane discussed earlier. G.P. Wells explains how these "Founding Fathers" pulled it off:

With the other three, Hogben discussed his idea for a new *Journal*. Crew at once declared that he had in hand enough cash from compensation for war [WW1] wounds to finance the initial project, and that he would be delighted to use it in this way. For other reasons, and not because of his generous offer, the other three prevailed on Crew to become the first Managing Editor … Crew persuaded Oliver & Boyd to print and publish the new Journal in Edinburgh. Hogben undertook the circularization of libraries, University departments, and Institutes. Finally, in October 1923, the first Part of the British Journal of Experimental Biology saw the light.

Initially, comparative animal physiology was underrepresented. In the first issue only one of the six articles fitted the bill, Maurice Yonge's comprehensive analysis of food capture and digestion in a clam as part of his PhD thesis (Yonge 1923). By the end of 1924 the contribution of comparative physiology to the journal picked up, and it included papers by Hogben and Carl Pantin. Hogben, true to his "enfant terrible" image, was soon to leave the UK because of fallouts with staff in Edinburgh and his inability to secure an academic position in England (Erlingsson 2016). In 1925 he managed to obtain a professorship at Montreal's McGill University. Carl Pantin then replaced him as secretary of the Society of Experimental Biology.

Although the *British Journal of Experimental Biology*'s beginnings seemed auspicious, it soon experienced financial problems. Subscriptions failed to accrue, as even some animal physiologists did not join in. In particular, as Erlingsson (2013) shows, Cambridge colleagues had just launched the *Biological Proceedings of the Cambridge Philosophical Society*, a periodical that gave all the appearances of treading on Hogben & Company's territory. But fortunately the Cambridge group relented and decided to morph their *Proceedings* into *Biological Reviews*, to which no original papers were invited. And to make matters even better, the *British Journal* found in George P. Bidder (1863–1954) a Cambridge zoologist and sponge expert of independent means, a financial supporter who saved the journal from insolvency by creating in 1925 the Company of Biologists to manage it. In 1930, since the journal had begun attracting foreign contributors, it dropped the *British* from its name.

As comparative physiology was slowly becoming institutionalized in Great Britain, a similar trend followed shortly in Germany. The man behind the creation of a German periodical devoted to animal physiology was no less than Karl von Frisch (1886–1982), renowned for deciphering the language of bees. He was born in Vienna to a closely knit Austrian family of distinguished professionals, scientists, and literary figures (Thorpe 1983). His mother, Marie Exner, was an artist, and his uncle, Sigmund Exner, a renowned physiologist who studied invertebrate vision and bird flight, among other topics. Frisch studied medicine at the University of Vienna, but under his uncle's influence he soon drifted to zoology. His uncle had produced an outstanding physiological study of the compound eyes of insects and crabs in which he gave an uncanny account of how insects see the world around them (Exner 1891). Frisch wrote in his memoirs (Frisch 1962) that Exner's monograph was "a comparative physiological work in the best sense of the word, at a time when this science as independent discipline did not yet exist." Frisch's first publications embroidered on his uncle's milestone work.

Frisch left Vienna to study zoology under Richard von Hertwig at the Zoological Institute of the University of Munich, where in 1910 he completed a doctoral dissertation on the nervous control of the fast skin-colour changes of fishes. He remained in Munich as a lecturer, making important research contributions. His research soon led him to the discovery of the role of the pineal gland in the skin-colour response of minnows to light and dark. There followed studies demonstrating that fishes can discriminate among colours. Then in 1913 he began his major research program on bees, which lasted for the rest of his life. By the time he founded the *Zeitschrift für vergleichende Physiologie*, he had produced papers or monographs on colour vision and shape recognition of bees (1915), their sense of smell in relation to the scent of flowers (1919), and their body language (1923).

Frisch was promoted to full professor in Munich in 1919, but in 1921 he moved to the University of Rostock and in 1923 to the University of Breslau. He returned permanently to Munich in 1925. It was during his tenure in Breslau that the idea of a journal dedicated to comparative physiology took shape. In his memoirs (1962) Frisch explained the predicament of the young zoologist yearning for comparative physiology in the early 1900s:

Physiology was almost exclusively associated with medicine. If animals were used for experiments in physiological institutes of medical schools, it was because you could not experiment on people. Studying frogs and rabbits helped one learn about the function of the human organs by analogy. To zoologists it made sense to be familiar with the wide Kingdom of animals from which to select models to establish a comparative physiology. But that did not happen. One was still too tied up with the morphological contemplation of the wealth of forms, which expeditions to distant lands and the exploration of the deep sea made even wealthier. In addition, one was not trained in physiological experimentation. So it was necessary for the zoologist-cum-physiologist to train with human physiologists at medical faculties, in such a way that the pioneers of this new branch of zoology would be equipped with the necessary methodological knowledge.

Like many of his generation, Frisch relied on Winterstein's *Handbuch* to gain appreciation for the breadth of knowledge accumulated so far. In addition, several colleagues who considered themselves comparative physiologists had joined him in publishing their research in the *Zoologische Jahrbücher*. Founded in 1886, the journal catered first to systematics and biogeography, but in 1910 it started a section to accommodate physiologists and more dynamic zoologists at large. Yet Frisch considered this a temporary arrangement; to him the scattering of publications dealing with comparative physiology in zoological journals or physiological journals of a paramedical nature were unsatisfactory. He turned to the publisher Ferdinand Springer to float the idea of a journal devoted entirely to comparative physiology (Frisch 1962).

By the time of Frisch's approach, the Springer publishing house already had a long history in Germany. Founded by Julius Springer (1817–1877) in 1842, it started by "publishing political periodicals, added children's books and schoolbooks to his programme, and then turned to forestry, science and pharmacy. [Julius] also published the work of some notable figures of German literature, such as Gotthelf and Fontane, and from the start encouraged publications on jurisprudence and economics" (Sarkowski 1996). When Julius's grandson Ferdinand (1881–1965) took over after World War I, Springer became the "leading German scientific publisher." Not only did Ferdinand accede to Frisch's request but he also published the Austrian's popular book *Aus dem Leben der Bienen* in 1927 [*The Dancing Bees*, 2016].

Karl von Frisch.
Courtesy of Universitätsarchiv München.

Frisch would later win the Nobel Prize for his landmark experimental study of bee behaviour.

To implement his vision of the *Zeitschrift für vergleichende Physiologie*, Frisch enlisted the assistance of a fellow zoologist who shared his interests. Alfred Kühn (1885–1968) was trained in physiology and zoology at the University of Freiburg, where he obtained his *Habilitation* in 1914 (Rheinberger 2000). After serving in World War I, he spent two years as an assistant to zoologist Karl Heider at the University of Berlin. In 1920 he was appointed head of the Zoological Institute at the University of Göttingen. Kühn's research interests focused first on embryology and cytology, but in Göttingen he turned to the comparative physiologies of animal orientation in relation to the sensory environment and of colour vision in bees (Kühn 1927). It is not clear if his fascination with bees was directly influenced by Frisch, but this confluence of interests certainly brought them into a close professional relationship.

If the goal of promoting comparative physiology as a discipline was shared by Hogben and his German counterparts, the strategy to achieve this goal in print could not have contrasted more. British comparative physiologists numbered too few to avoid compromise with other experimental zoologists and even botanists for the foundation of the *JEB*. The Germans had no such qualms. The number of German comparative physiologists had reached a critical threshold even in the absence of a journal outlet. So Frisch and Kühn signalled their determination to exclude any experimental zoology other than comparative physiology from the journal by encrypting "comparative physiology" in its name.

But what is in a name? Frisch and Kühn made sure that the name of their journal was not an empty promise. Perusal of the journal from the very first issue in the spring of 1924 shows that nearly all the articles dealt with one comparative physiology problem or another. Reflecting the founders' interests, sensory physiology in a variety of lower vertebrates and invertebrates dominates early on. Vision, the sense of smell and taste, hearing – even sensitivity to temperature – are investigated. Swimming and other locomotion modes are examined along with muscle physiology. Frisch and Kühn of course contributed their own articles. Even Hans Winterstein registered his approval by submitting articles to the journal.

What is particularly striking in the approach of the contributors is their departure from the comparative model of many of their predecessors such as Loeb, who looked for basic, common physico-chemical processes in the variety of studied animals. The new generation represented in the *Zeitschrift für vergleichende Physiologie* seemed united in their desire to reveal the unique physiological solutions of individual (sometimes exotic) animal species to the life challenges they specifically faced. In a sense these physiologists used their newly promoted discipline to showcase the diversity of life, to celebrate biodiversity by deeds, if not consciously by name. For budding zoologists in other countries as well as in Germany, this turn was exciting and motivating. Talbot H. Waterman (1914–2010), a Yale University physiologist best known for his research on how aquatic animals use polarized light for navigation, recalled "the intellectual excitement I felt as a student on 'discovering' the Zeitschrift fur Vergleichende Physiologie … Somehow these publications seemed to focus and reinforce my budding ambition to become a biologist. No doubt this was because they aroused great expectations for what could be accomplished by the comparative approach to experimental zoology" (Waterman 1975).

∼

Another great source of inspiration in addition to the *Zeitschrift* was a Danish physiologist whose ongoing career was pivotal for the development of animal physiology. Not only was August Krogh (1874–1949) an exemplar of what Karl von Frisch had said about the need to ground comparative physiology in medically oriented physiological methodology, but Krogh was also a fount of wisdom about the relevance of comparative physiology early in the twentieth century. Krogh's life and achievements are amply documented in a book by his youngest daughter, Bodil Schmidt-Nielsen (1995), and the following account borrows amply from it.

August Krogh was born in a small Jutland town to the family of a brewer. He showed no enthusiasm for school – paying more attention to the life of insects in the wild than to class instruction – and he even abandoned school temporarily at the age of fifteen to join the Danish Navy. In 1893 he enrolled at the University of Copenhagen intending to study physics, but soon, under the influence of the zoologist William Sørensen, whom he befriended and who mentored him, he switched to zoology. Zoology was a call he could not ignore, he wrote to his father, who thought there was no future for his son in that field (B. Schmidt-Nielsen 1995). But a casual suggestion by Sørensen for Krogh to attend lectures by the physiologist Christian Bohr at the Medical School completely turned his life around.

Bohr's outstanding research on the properties of blood and on respiration bewitched Krogh, who pleaded with Bohr in 1897 to work under him in the Physiological Laboratory. But Krogh, unlike other physiology students of the day, had no interest in pursuing a medical degree and instead completed a master's degree in zoology in 1899. In 1898 he was officially appointed Bohr's assistant and his research on the cutaneous and lung respiration of frogs served as the basis for his PhD dissertation in 1903. During these years Krogh started to display his superb, lifelong gifts for designing and building apparatus tuned to answer specific physiological queries with the greatest accuracy.

As Bohr's assistant, Krogh participated in a large-scale project on respiratory gas exchanges in the skin and lungs of snail, fishes, and tadpoles. There followed studies that "explained the various mechanisms used by animals to keep themselves afloat in the water without active movements [gas-filled floaters such as swimbladders]. It was the first Krogh presented of what is now called comparative physiologic studies" (B. Schmidt-Nielsen 1995). In fact, early on Krogh used the word first coined by the French: "zoophysiology." In his 1904 paper titled "The tension of carbonic acid in natural waters and especially in the sea," his theoretical considerations and calculations of production and removal of CO_2 from the earth's atmosphere led him to conclude that the burning of coal was a major factor in what we call today the greenhouse effect. He also concluded that the sea acted to regulate atmospheric CO_2 by absorbing it. Also in 1904 Krogh and Bohr discovered how CO_2 displaces oxygen from the blood haemoglobin to make it available to tissues. It became known as the Bohr effect, even though the paper reporting this important finding was coauthored with Krogh, who invented and constructed the apparatus to discover and measure it (B. Schmidt-Nielsen 1995).

Krogh's relationship with Bohr was occasionally tested by issues related to publication. Already in 1898 what was planned as their first publication together unravelled over the choice of language. Bohr studied under German

August Krogh.
From photo archives of the Marine Biological Laboratory, Woods Hole.

physiologists and it came naturally for him to publish in German, which, after all, held sway among continental European scientists. But Krogh, who had made a conscious effort early in his life to master English, saw things differently. The crux of the matter was the anti-German sentiment of many Danes, including Krogh's parents, ever since Bismarck had engineered Prussia's takeover of the Danish province of Schleswig-Holstein. In a letter to Bohr, Krogh explained his position in guarded language that reflected on his character:

> Honorable Professor Bohr, Following careful consideration I ask you not to include my name as coauthor on the investigation of the lung and skin respiration in frogs. What brings me to this is first of all, that all along and especially now that the publication is at hand, I have felt that my part of the work aside from the purely mechanical experimentation is too small to warrant that I become coauthor. I shall be totally satisfied with being acknowledged as the one who carried out the experiments. Furthermore, I must add that I believe that my parents would not be pleased were my first publication to be published in German. (cited by B. Schmidt-Nielsen 1995)

Krogh soon relented, however, probably having realized that his position on principle would become untenable for his career. Yet he took every opportunity to publish an English version of papers originally released in German or Danish. Krogh finally gained his independence as a researcher with the building of his own Zoophysiological Institute in 1910, where he and his wife, Marie, herself a medical doctor with outstanding research talents, teamed up for a decade of remarkable research output climaxing with Krogh's Nobel Prize in Physiology or Medicine in 1920. The Nobel Prize rewarded research showing how capillaries play their role in delivering oxygen to the tissues. It was medically oriented work, but it is important to remember that Krogh was no medical doctor and that his institute was affiliated with the Science Faculty, not the Medical School. He was the first comparative physiologist to win the Nobel.

Among other noteworthy "zoophysiological" achievements was the game-changing research that he conducted between 1910 and 1915 on how the ingeniously designed respiratory system of insects copes with oxygen demands at rest and during flight. He also showed how aquatic insects collect air bubbles and use the bubbles' oxygen for respiration under water. And, late in life, he and his bright student Torken Weis-Fogh used air tunnels to study particularly the role of ambient temperature on insect flight performance. They found that the air in sacs attached to the insect tracheal system was pushed out to the tissues by the insects' wing-muscle movements during flight, a system reminiscent of ventilating bellows in the old-fashioned blacksmith forge. Even in old age, in the late 1940s, Krogh's fascination with insects, dating back to his childhood, knew no bounds.

Many documents written by Krogh have survived to show how his mind worked when designing experiments or suggesting how the scientific method should be used. His was a restless mind and he explained how sleep time helped him think about his work:

> A considerable part of my work was done in bed during the night when I would try to visualize the processes studied and the experiments to be carried out. I found that I could visualize fairly complicated apparatus and all details of their working. The constructive ideas would come, apparently, out of nowhere, but the visionary examination of them was a conscious and rational affair. I never made, and even now never make, drawings, not even rough sketches, until the construction of an apparatus was complete, because I found that a drawing would hamper the free flow of ideas and bind me down to that particular solution of the problem. (Krogh 1938, quoted by B. Schmidt-Nielsen 1995)

In a remarkable text on the progress of physiology current as of 1929, Krogh did not spare his criticism of what he considered abuses of the experimental method in his days. "Too many experiments," he wrote, "are done and too few thoughts are bestowed upon them." And his daughter added: "He himself was sparing in the number of experiments he carried out, and he found it useless when investigators did fifty identical experiments showing exactly the same phenomenon, where fewer more crucial experiments under different experimental conditions would have told a complete story" (B. Schmidt-Nielsen 1995). The exquisite accuracy of his measurements, thanks to his wizardry in designing apparatus, obviated the need to endlessly repeat experiments.

In the same text (Krogh 1929) he forcefully promoted the institutionalization of comparative physiology, clamouring for "the creation in the science schools and in close cooperation with the departments of zoology of
chairs and laboratories of comparative physiology, animal physiology or
zoophysiology. The name does not matter much, though I confess that there
is one name with which I have no sympathy, – that of 'general' physiology."
Such a pronouncement by a highly influential physiologist relegated Loeb's
concept to the rearguard of physiology. Krogh added that "the route by which
we can strive toward the ideal is by a study of the vital functions in all their
aspects throughout the myriads of organisms."

One formulation for which Krogh is particularly famous is what the great
biochemist Hans Krebs (1975) called the Krogh principle. Krogh explained
himself in his 1929 paper:

> For a large number of problems there will be some animal of choice
> or a few such animals on which it can be most conveniently studied.
> Many years ago when my teacher, Christian Bohr, was interested in the
> respiratory mechanism of the lung and devised the method of studying
> the [gas] exchange through each lung separately, he found that a cer-
> tain kind of tortoise possessed a trachea dividing into the main bronchi
> high up in the neck, and we used to say as a laboratory joke that this
> animal had been created expressly for the purpose of respiratory phys-
> iology. I have no doubt that there is quite a number of animals which
> are similarly "created" for special physiological purposes, but I am
> afraid that most of them are unknown to the man for whom they were
> "created" and we must apply to the zoologists to find them and lay our
> hands on them.

Although this utilitarian application of comparative physiology is com-
mendable, Krogh expressed an aesthetic fondness for comparative physiology when he said: "I want to say a word for the study of comparative
physiology also for its own sake. You will find in lower animals mechanisms
of exquisite beauty and the most surprising character" (Krogh 1929). The
beauty and surprise were embodied in his research on the osmotic mechanisms that animals use when their environment changes, using examples

such as eels migrating from freshwater to sea water for breeding or the amazing diversity of respiratory mechanisms found throughout the animal kingdom. As Steven Vogel (2008), the great specialist of comparative biomechanics, put it, Krogh's laboratory "had a consistently biological orientation, focusing on the general physiological problems of animals, with non-human material serving as far more than experimentally convenient surrogates for ourselves."

All Krogh's hopes for the growth of comparative physiology expressed on the eve of the 1930s materialized in the next two decades. New cohorts heeded Krogh's ideal for comparative physiology, and a budding branch – comparative biochemistry – asserted its place in this grand scheme. Particularly gratifying for Krogh near the end of his life was witnessing the embodiment of his legacy in the budding career of his daughter Bodil and his son-in-law Knut Schmidt-Nielsen.

5

American Schools of Comparative Physiology

ﾑ

Comparative animal physiology integrates and coordinates functional
relationships which transcend special groups of animals. It is concerned
with the ways in which diverse organisms perform similar functions.
~ C. Ladd Prosser (1950)

If comparative physiology became consciously acknowledged in so many
words by August Krogh's generation and had earned its place as a sub-
discipline of zoology, it was not until the decades between the 1930s and 1950s
that it began to achieve its independence and make its mark. Notwithstanding the early contributions by Great Britain and Germany spelled out in the
previous chapter, the greater part of the subsequent surge in this field originates in the United States and, as discussed in chapter 6, in the Belgian
school of comparative physiology and biochemistry. We will now examine
the development and impact of these schools.

In the United States two giant figures of contrasting personality and approach to animal physiology were to dominate the field: C. Ladd Prosser and
August Krogh's son-in-law Knut Schmidt-Nielsen. Both spawned outstanding followers of their respective approaches.

But preceding these two luminaries, the unsung pioneer of comparative
animal physiology in the United States must certainly be Charles Gardner
Rogers (1875–1950). Whether there was a conspiracy of silence around him
or whether his contribution simply went under the radar – he is almost never
mentioned by future American leaders in the field – this simple fact cannot

be dodged: he was the first appointed professor of comparative physiology in America. Born in upstate New York, Rogers graduated from nearby Syracuse University and earned his PhD in 1904 under the supervision of Jacques Loeb at Berkeley (McEwen 1951). He went on to hold a professorship back at Syracuse. In 1913 he was appointed professor of zoology at Oberlin College in Ohio, where he remained until his death. From 1915 until his retirement in 1941 he held the chair of comparative physiology at Oberlin.

Rogers usually spent his summers at the Marine Biological Laboratory in Woods Hole, Massachusetts, where he conducted some of his physiological researches on local marine animals. Notably, he showed how cold-blooded animals adapted their body temperature to ambient temperatures in their surroundings (Rogers and Lewis 1916). He also wrote the first American textbook of comparative physiology (Rogers 1927). In the book's preface Rogers put the subject matter clearly in context: "[The] physiology of animals is really functional zoology ... Since it is functional zoology, it concerns itself with the primary functions of animals of all groups, especially of the invertebrates which constitute probably not far from 93 percent of all known species of animals." Interestingly, he emphasized the "conception of evolutionary changes in animal functions" and the "physiological bases of animal relationship." The book's popularity as a course textbook was such that a second, enlarged edition appeared in 1938. Nevertheless, perhaps because of his academic affiliation with Jacques Loeb, then considered a general physiologist, or because of his professorship in a small college with little opportunity to train graduate students – or his admittedly lesser intellectual stature – Rogers's reputation was limited among the up-and-coming animal physiologists of a comparable bent.

One of the latter was Clifford Ladd Prosser (1907–2002) who, ironically, was born in the Genesee Valley near Rochester, a mere 120 kilometres west of Rogers's birthplace. "Ladd" Prosser is rightly regarded as a giant in the field and, as fellow practitioner George Somero (2009) writes, "a principal catalyst in the development of the broad field of comparative physiology." Prosser was an enabler not only by virtue of his contagious curiosity about nature's treasures, which "led him to ask penetrating questions that continue to challenge and motivate us," but also through his ability to situate various aspects of the field in insightful review articles. More significantly, as Somero makes plain, "he helped to refine a philosophical context – the comparative method – that has enabled biologists to exploit the diversity of nature to elucidate the common, basic principles that characterize living systems."

Prosser's future as a zoologist and comparative physiologist was rooted in his childhood and early adulthood. Late in his life Prosser himself (1986) reflected on his coming-of-age:

> I spent my youth in Avon, a small town twenty miles south of Rochester, New York. There was frequent electric train service to the city, and opportunity to hike along the Genesee River and up and down the hills leading to western Finger Lakes and their outlets. My interest in nature was stimulated by weekly hikes with my father. Fossils were abundant in the slate and shale lining the many gullies. I started collecting insects with two pals with Blatchley's *Coleoptera*, and plants with Gray's *Botany* in the seventh grade. I majored in biology at the University of Rochester where a turning point in my career was a course in Physiological Psychology in which the text was Herrick's *Neurological Foundations of Behavior* [see chapter 9]. During this course I decided that the neural basis of behavior could better be studied with invertebrates than with mammals. I went to The Johns Hopkins University for graduate study, and wrote a thesis on the physiology of the nervous system of earthworms. The summer after my first year there was spent as a research assistant to S.O. Mast at the Mount Desert Biological Station [Maine]. My first published paper was on amoeboid movement. During that summer (1930) I was out at every low tide becoming acquainted with invertebrate animals.

Prosser's doctoral program was conducted under the aegis of Samuel O. Mast (1871–1947) at the Zoological Laboratory of Johns Hopkins in Baltimore. Mast was a specialist in unicellular organisms, and especially their reactions to light; he was among the first foreigners to publish (in English) in Frisch and Kuhn's *Zeitschrift für vergleichende Physiologie*. Prosser's first scientific article was co-authored with his mentor and concerned the effect of various physical and chemical factors on the dynamics of locomotion in *Amoeba* (Mast and Prosser 1932). By strange coincidence, the article appeared in the first volume of a new journal – the *Journal of Cellular and Comparative Physiology* – in which comparative physiology was meant to loom large next

C. Ladd Prosser at the Marine Biological Laboratory, Woods Hole, in the 1930s.
MBL Still Image: "C. Ladd Prosser," 2012-06-11.

to the physiology of unicellulars. The content of the first volume raised great hopes, as the majority of papers had a comparative physiology flavour. Sadly, however, the flavour gradually dissipated, and "comparative physiology" was dropped from the journal's name thirty some years later.

Even though Prosser cut his research teeth on Mast's animal model, for his PhD thesis he decided to follow his own instinct and investigate the nervous system of earthworms. His approach was rather conventional – to follow the development of the earthworm's behaviour in relation to the maturation of the structure of nervous system components (Prosser 1933a), or to look for the seat of responses to light in the nervous system by selective cuts or ablations (Prosser 1933b). Much more original was the postdoctoral research that he conducted in the laboratory of Hallowell Davis in the Department of Physiology of Harvard Medical School. There Prosser met the challenge of learning the daunting technique of recording nerve electrical potentials and applying it to the nervous system of crayfish. As the introduction in the first paper on this work reveals, Prosser (1934a) realized early on how important it was to follow (unwittingly) Krogh's principle: "For many problems there is an animal on which it can be most conveniently studied":

The principal difficulty encountered in studying electrically the interaction between the central neurons is the confusion resulting from the large number of neurons in any given region of the central nervous system. Characteristic of invertebrate nervous systems is their more diffuse nature and the fact that each ganglionic center contains relatively few cells, when compared, for example, with one segment of the spinal cord of the cat. Very few attempts have been made to apply to invertebrate ganglionated systems the electrical methods used in studying the peripheral elements in classical vertebrate preparations. The present series of papers is the result of an effort to elucidate some of the problems of the central nervous system by studying action potentials in the relatively simple nervous system of the crayfish.

Prosser's reasoning led to important discoveries. First, he found that brain neurons fired action potentials spontaneously and rhythmically, without stimulation from sensory input (Prosser 1934a). As he later wrote in his 1986 autobiography: "This discovery dealt a blow to the behaviorist dictum that all patterned behavior must be initiated by sensory input." We know now that this phenomenon forms the basis of the motor programs (or brain algorithms, so to speak) that manage automatic animal activities such as locomotion or breathing. Second, he made the astonishing discovery of neurons sensitive to light in the tail-end ganglion of the crayfish (Prosser 1934b). Such neurons are now believed to help synchronize locomotion during the daily light-dark cycle. The latter represents the type of exotic discovery unveiled when the curiosity of the prepared mind gets to work.

A key asset that contributed to Prosser's growth as a scientist was his gregariousness; he easily made contacts with established and renowned physiologists and did not fail to learn from them. It happened at Johns Hopkins, at Harvard, and during summers at the Marine Biological Laboratory (MBL) in Woods Hole. Although, in contrast to Naples, the MBL had neglected animal physiology since its foundation in 1888, physiologists started to flock there for summer research in the late 1920s and 1930s. At the Cape Cod facility they found the "most convenient" animal models for the questions they asked; the horseshoe crab, for instance, for understanding visual mechanisms at the cellular level. Prosser made good use of the MBL during his training years and throughout his career, not only for research but also to engage with colleagues and students alike.

The lack of opportunity for academic employment during the Great Depression forced Prosser to mark time in England, where he "went to work [at Cambridge] with Professor E.D. Adrian, who had recently published on electrical activity in insect nervous systems." Back to America in 1934 he had to work as an assistant in a laboratory at Clark University. Finally, thanks to a recommendation by Harvard zoophysiologist George H. Parker – one of the academic lights he "cultivated" – Prosser landed an assistant professorship at the University of Illinois at Urbana-Champaign in 1939. After the hiatus of World War II, during which he was enlisted in warfare research, including a stint as manager at the Metallurgical Laboratory of the Manhattan Project, his career took off in earnest.

In the 1940s, Prosser diversified his research interests to range from the physiology of smooth (non-skeletal) muscles in various invertebrates to the biochemical mechanisms that fishes use to acclimate to cold and warm environments (Somero 1986). And then in 1950 his landmark book of comparative physiology was published, catapulting him to the top of his discipline.

In his autobiography he gave an account of the genesis of the book as well as what the discipline meant to him and his generation:

> While at Clark and in my first two years at Illinois, I developed a plan for a book on comparative physiology which would emphasize evolutionary and ecological applications of physiology. The idea of such a book appealed to the W.B. Saunders editor, and the first edition of *Comparative Animal Physiology* appeared in 1951 [marked as 1950 on the copyright page]. I had several collaborators and the book set the tone for comparative physiology for many years. A second edition came out in 1961 and a third edition in 1973. I enjoyed treating adaptations of animals to various environments. Comparative physiology differs from other kinds of physiology in that the comparative approach uses the kind of organism as an experimental variable, and it emphasizes the long evolutionary history leading to life in diverse environments.

None of Prosser's predecessors had so clearly articulated the mission of comparative physiology around ecological and evolutionary themes. He was defining a new paradigm for approaching the study of animal functions. It is worth enlarging on his philosophy regarding his field, as he explained it at greater length in the book preface:

> The objectives of comparative physiology are: (1) to describe the diverse ways in which different kinds of animals meet their functional requirements; (2) to elucidate evolutionary relationships of animals by comparing physiological and biochemical characteristics; (3) to provide the physiological basis of ecology, describing the mechanisms of tolerance of the stresses of particular habitats and the functional adaptations underlying extension of the range of a population; (4) to call attention to animal preparations particularly suitable for demonstrating specific functions; and (5) to lead to broad biological generalizations arising from the use of kind of animal as one experimental variable.

Prosser also stressed the importance of the notion of adaptation. In the introduction to *Comparative Animal Physiology* he wrote:

Foremost among general principles which emerge from a study of comparative physiology is the functional adaptation of organisms to their environment. The distribution of a species is determined through natural selection by its limits of tolerance. Every species can live within certain limits of variation of each environmental factor. One environmental factor may limit the distribution of one group, and another may limit another group. Salinity of an aquatic habitat limits some animals, oxygen tension limits some, and temperature extremes limit others. Explanations of both restricted distribution and widely diversified distribution can be obtained by examining physiological reactions to environmental stress.

And he added that any physiological response that allows survival is adaptive. With this Darwinian note, Prosser made clear that comparative physiology should put its shoulder to the wheel with other zoological disciplines and genetics to uncover features of evolutionary significance. This he attempted in his book with the assistance of a few collaborators, who wrote or co-wrote eight of the twenty-three chapters. Each chapter covers a specific function; water, nutrition, respiration and metabolism, photoreception, muscles and electric organs, and so forth. The compartmentalization of function by chapter gives little attention to the systems approach, which would surface in later editions (1973 and 1991).

In acknowledging his predecessors, Prosser made no mention of Charles Rogers's book as a foundational text, but instead stated: "[T]he only truly comprehensive account in recent years is Buddenbrock's *Grundriss der vergleichenden Physiologie.*" Wolfgang von Buddenbrock (1884–1964) took a backseat to Karl von Frisch, August Krogh, and others insofar as European reputation in comparative physiology is concerned. Born into a Prussian dynasty of military officers and landowners (*Junkers*), Buddenbrock was a late bloomer who lost his way in engineering studies before switching to zoology (Schaller 1985). He studied for a year in Jena with Ernst Haeckel, who sent him to Messina (Sicily) to collect marine specimens. He then moved to Heidelberg, where he studied under the celebrated zoologist Otto Buschli and obtained his PhD in 1910. After taking an assistantship at the Zoological Institute of the University of Berlin, he was appointed professor at the University of Kiel, where he remained from 1923 to 1935. He published extensively on sensory physiology as related to the orientation of marine animals in their environment, and on the "love life" of animals.

It was during his tenure in Kiel that Buddenbrock published his *Foundation of Comparative Physiology* (1928), of which a second edition appeared in 1938. Unlike Winterstein's *Handbuch*, which targetted scholars and research students, Buddenbrock's *Grundriss* is considered the first German textbook of comparative physiology intended for classroom use. While the likes of Frisch studied one species at a time (bees, fishes), Buddenbrock tended to compare a particular physiological feature among different species within a single original paper, and in this way he was considered a deep-dyed comparative physiologist (Bückmann 1985). Even Prosser did not push his practice of comparative physiology this far; his comparisons turned up in synthetic review articles as well as in his textbook.

While the impact of Buddenbrock's *Grundriss* was limited in time and space – it was influential mostly in German-speaking countries in the decades between 1930 and 1960 – Prosser's *Comparative Animal Physiology* remains a classic in the field through its re-editions. Prosser's influence was twofold. On the one hand his textbook made its presence felt in the curriculum of universities and colleges in the United States and elsewhere just as comparative physiology was making its entry as a class subject and in research labs. As Greenberg and collaborators (1975) remarked, "It had no serious competitors and few imitators." What is more, Prosser himself, through his own contacts with students in his classroom or in his research lab, influenced many aspiring animal physiologists to follow in his footsteps. Somero (2009) writes of Prosser's "success in motivating and energizing a large cadre of young scientists, many of whom [went] on to become leaders in their fields." Somero attributes the fascination that Prosser exercised on generations of students over seven decades to his character and particular talents:

> He not only was incessantly curious about nature but he also wanted to know what you knew or thought about a myriad of different topics. This desire to learn from others was coupled with capacities for filing away all that he had learned and being able to integrate and synthesize this information both horizontally among disciplines and vertically along the reductionist-holistic axis. Having grown up with many of the

fields of physiology and, in some cases, serving as the originator of these fields, Ladd was successful in keeping up on the literature in a way that would be impossible in this day of fragmentation of knowledge and overwhelming output of papers.

In his later years Prosser incorporated his concept of adaptation in an ambitious book, *Adaptational Biology* (1986), which went beyond comparative physiology and adopted a multidisciplinary, more holistic approach to the topic. Already in 1969 Prosser had distanced himself from approaching adaptation through the lens of the animal physiologist alone and embraced it instead from the standpoint of the biologist at large (Prosser 1969). But this standpoint introduced a difficulty: "If I were to ask 100 biologists for the meaning of adaptation," he wrote, "I might get 100 different definitions." Prosser was not deterred by the challenge and firmly proposed his own definition:

Adaptation refers to any property of an organism which permits physiological activity and survival in a specific environment; adaptation is characteristically related to stressful components of the environment although it may relate equally well to a total environment. Adaptive characters have genetic basis but may be expressed according to environmental needs.

Further into the 1969 article, Prosser linked adaptation with environment and evolution:

Adaptive variations may be measured in individual populations, or higher taxonomic categories. They include anatomic, physiologic, and biochemical characteristics of individual organisms which relate these individuals adaptively to a specific environment. In an evolutionary sense, only those variations that arc adaptive are retained. Natural selection is the only known mechanism for fixation of adaptive variations and forms the basis for speciation.

Such reflections led Prosser to formulate a subfield of animal physiology that he called "environmental physiology," which was to figure prominently in the structure of the 1991 revision of his *Comparative Animal Physiology*. He used a concrete example to make his point:

> A physiological concept of biological species may be derived as follows: If no two species can occupy the same ecological niche or the same geographic range throughout their life cycles, it follows that every species must be uniquely adapted to its particular niche and range. Hence, if we could quantitatively describe the physiological adaptedness of a species to its ecological niche and geographic range, we would have a truly meaningful description of the species. One of the goals of environmental physiology is to achieve some understanding of the molecular basis for natural selection.

The goal Prosser alluded to was on the road to achievement in the 1960s through the development of the field of comparative biochemistry, of which more later. But for now, in the context of his exposé, "adaptive variations [were] considered for individuals, not species." If the role of physiological adaptations to the environment in the formation of species and in evolutionary processes at large defined what Prosser was after, for his great rival in forging modern animal physiology the most meaningful things were those that happened to individuals. Knut Schmidt-Nielsen (1915–2007) differed from Prosser not only in personality but also in philosophical outlook. While Prosser thought of himself as a zoologist who happened to look at animals from the physiological angle, Schmidt-Nielsen went the full distance, excusing himself from the ranks of zoologists to be labelled solely as a physiologist.

~

Knut Schmidt-Nielsen's memoirs, published in 1998, provide details of his personal as well as his professional life. These, together with the appraisal of Schmidt-Nielsen's life by Steven Vogel (2008), his successor as James B. Duke Professor of Physiology at Duke University, present ample information upon which to base the following account.

Knut Schmidt-Nielsen was born in Trondheim, Norway, to highly intelligent, university-educated parents. To say that Knut had an auspicious start in life is an understatement. His father earned his doctoral degree working in the lab of German chemist Eduard Büchner, who won the Nobel Prize in 1907. His Swedish mother earned her own PhD under the supervision of the famous physical chemist Svante Arrhenius, a 1903 Nobel recipient. But in spite of this impressive pedigree, Knut disliked school and fared poorly as a student. In the opening sentences of his memoirs, he quipped: "It has been said that the primary function of schools is to impart enough facts to make children stop asking questions. Some, with whom the schools do not succeed, become scientists." In a nutshell, he underscores the unrelenting curiosity about nature that drove his career, never ceasing to ask questions and strive for answers.

At first Knut contemplated engineering for his university studies, but he soon changed to zoology at the University of Oslo. Interestingly, he was known to often take an engineering approach to physiological questions about animal life. During his studies he came upon a copy of Richard Hesse's *Ecological Animal Geography* (1924), which greatly influenced him. Hesse (1868–1944) had started his career as a conventional zoologist and comparative anatomist, and only later broke new ground with his zoogeography book, which contributed enormously to the emerging field of ecology. "Ecological zoogeography," Hesse wrote, "views animals in their dependence on the conditions of their native regions, in their adaptation to their surroundings, without reference to the geographic location of this region, whether in America or Africa, the northern or the southern hemisphere." This meant, for instance, that the desert environment itself was the focus, not geographical location of that desert. Schmidt-Nielsen acknowledged his debt to Hesse in his memoirs: "Hesse's book brought to my attention questions that have occupied much of my scientific life: What physiological characteristics permit some animals to live in environments that to others are hostile and uninhabitable?"

When Knut told his father of his desire to pursue physiological studies, his father wrote to August Krogh asking if he could find room for his son in his laboratory. Krogh agreed and in the fall of 1937 Knut went to Copenhagen. There he was assigned to compare water and salt regulation between freshwater (crayfish) and marine (crab) crustaceans. He summarized his main finding thus: "I found that in brackish water the crab can maintain higher [salt] blood concentrations, but in very dilute sea water or in fresh water it cannot and dies" (Schmidt-Nielsen 1998). By mid-1938 Knut had fallen for Krogh's youngest daughter, Bodil, then a nineteen-year-old studying dentistry. They married in September 1939 but six months later found themselves trapped in Copenhagen by Germany's invasion of Denmark and Norway. Despite wartime shortages and deprivations, Knut managed to complete a doctoral thesis on how fatty acids are absorbed by the intestine, and Bodil conducted important research on the formation of dental caries. In addition to their professional work they produced two children, Astrid and Bent.

In 1946 a visit to Scandinavia by the American physiologist Lawrence Irving (1895–1979) and Swedish-born Per Scholander (1905–1980), both stationed at Swarthmore College, sealed the destiny of the Schmidt-Nielsens. The visit led to an invitation to the couple to fill positions of research associates at Swarthmore. With dire conditions for academic research prevailing in postwar Scandinavia, Knut and Bodil had little choice but to accept. But before we go into their initial years in America, it is worth visiting the careers of Irving and Scholander, both outstanding animal physiologists in their own right – the latter Knut Schmidt-Nielsen's avowed scientific hero.

It can be said of Lawrence Irving that he rendered great service to American animal physiology not only through his own pioneering research but also through his agency in facilitating the moves of Scholander and the Schmidt-Nielsens to the United States. Born in Boston, Irving completed a master's degree at Harvard in 1917 and entered military service as an infantry lieutenant in the American Expeditionary Force and Army of Occupation during and shortly after World War I (Dawson 2007). When he resumed academic pursuits he completed a PhD at Stanford University in 1924 on the biology of starfish. Like many of his generation he undertook postdoctoral studies in Germany, in his case under Gustav Embden at the Physiological Institute in Frankfurt. Embden was what was then called a "physiological chemist" who was renowned for his research on metabolism. He demonstrated the central role of the liver in the overall metabolism of the body and traced all the steps from the breakdown of glycogen to the formation of lactic acid in muscle. As a result of Irving's research fellowship in Frankfurt, he produced papers on muscle chemistry for years thereafter (Dawson 2007).

In 1927 Irving was appointed associate professor at the Department of Physiology of the University of Toronto. It was during his decade-long tenure

Bodil Krogh Schmidt-Nielsen (left), Knut Schmidt-Nielsen (centre), and assistant Barbara Wagner (right). From Bodil Schmidt-Nielsen (1995). Access number 90-105 — Science Service, Records, 1920s—1970s, Smithsonian Institution Archives.

in Toronto that he embarked on the program of physiological research that singles him out as the forefather of the kind of comparative physiology later pursued by the Schmidt-Nielsens and others. "Expeditionary physiology," as Irving labelled his innovative approach, was predicated on conducting primary studies in the field and supplemental studies in the lab. The beaver, Canada's emblematic animal, was among the first to be subjected to the new approach (Irving and Orr 1935; Irving 1937). Irving went to Algonquin Park north of Toronto and showed how the beaver could dive for up to fifteen minutes and adapt by slowing its heart rate and increasing blood flow to the brain while reducing it to the muscles. The other animal of special interest to him was the harbour seal. For this project Irving visited the St Andrews Biological Station in the province of New Brunswick (now the Huntsman

Marine Science Centre). Irving and his team found that the seal's response to apnea differed from the beaver's only in scale not in kind (Irving et al. 1935a,b).

When he was appointed chair of the Department of Zoology at Swarthmore College in Pennsylvania in 1937, Irving returned to the United States. He had established bonds with August Krogh, and the famous Danish physiologist travelled to Swarthmore to deliver lectures that formed the basis of his 1941 book on respiratory physiology. It was through Irving's personal relationship with Krogh that his lab was able to add Per Scholander to its roster in 1939, beginning the legendary research program on diving physiology by the Irving-Scholander team.

Per Scholander was born in Sweden to an engineer father and a Norwegian-born mother, a musician (Schmidt-Nielsen 1987). In his early years Per displayed traits reminiscent of those previously mentioned for Prosser and Schmidt-Nielsen. As he himself admitted: "There is one thing I realized very early. I had a curiosity that craved research of any kind, and could not think of anything else. I was deadly afraid of winding up as a school teacher and decided that the surest way of avoiding that would be to go into medicine" (Scholander 1978). After his parents' divorce, he and his mother moved to Norway, where he started medical studies at the University of Oslo. Bored with his classes, Scholander started collecting lichens. His resulting interest in plants led to his tagging along on official expeditions to Greenland and Spitzbergen, thanks to the sponsorship of a botanist at Oslo University. He managed to earn a doctoral degree in botany along with one in medicine.

The seals and diving sea birds that Scholander saw on his trips to Greenland and Spitzbergen aroused his curiosity and he characteristically went to the core of the physiological puzzle of diving (Schmidt-Nielsen 1987): "He clearly saw that many important questions needed answers, such as 'How do diving seals get enough oxygen?' and 'Why don't they get divers' disease as humans do after descending to similar depths?'" With no academic status, he improvised a lab at the Physiological Institute of Oslo University and built his own equipment with materials at hand in order to answer these questions. In the seminal monograph that resulted from his labour (Scholander 1940), he confirmed Irving's findings but added a key element to account for the diving performance of seals. Seals can hold a lot more oxygen in their blood than humans and their total blood volume is also larger than in humans, so they can remain submerged longer, relying on their extra store of oxygen. Scholander's monograph, Schmidt-Nielsen asserted, "remains the foundation for what we understand today of the physiology of diving animals."

Scholander's early work, especially the methodological wizardry that filtered through his preliminary papers, so impressed Krogh that he invited Scholander to give a lecture on his work in Copenhagen. Knut Schmidt-Nielsen, who was then Krogh's student, recalled the lecture in 1987: "I sat there completely spellbound by his brilliant presentation and the simple and logical answers he provided to questions that long had puzzled physiologists who contemplated the mysteries of diving physiology." Krogh, realizing the close relationship between Scholander's research and Irving's, arranged a Rockefeller fellowship for Irving so that the pair could team up in Swarthmore, safe from the anticipated disruptions of the impending war. Irving and Scholander continued to make contributions to diving physiology in a variety of mammals and birds until the United States joined the war effort in 1941.

After the war Irving and Scholander asked a new set of questions: How do animals living in arctic climates keep warm and survive? With the help of the US Navy they set up a research laboratory at Point Barrow in Alaska, what Scholander (1978) described as "a quonset hut well insulated and heated for exacting physiological work." He reported their findings with his trademark simplicity and directness:

Our work on warm-blooded animals concentrated on measurements of insulation, metabolic rate, and body surface temperatures. It turned out that mammals larger than the fox could sleep with basal oxygen consumption at temperatures as low as −30° to −40°C, which were the coldest we could produce. The smaller ones, like weasels or lemmings, started to shiver at + 15°C; below that, they increased their heat production essentially proportional to the deviation from their internal temperature of + 37°C, as would be expected from Newton's law of cooling … It goes without saying that the animals adjusted their insulation by raising or lowering furs or feathers, curling up with the nose under the tail, etc. In the appendages of aquatic animals (seals, whales, water fowl), arteriovenous countercurrent systems are common heat savers.

Scholander and Irving also found amazing adaptations to freezing in cold-blooded animals; that some fishes from Labrador, for instance, have the ability to survive in water supercooled to $-3°C$ as long as they can avoid contact with ice crystals; it was later found that Antarctic fishes can survive even in contact with ice crystals because their blood contains a special protein that acts like an antifreeze (DeVries and Wohlschlag 1969). In the same period Scholander solved "the intriguing problems of how gases are secreted into the swimbladder of fish" (Schmidt-Nielsen (1987). This is but a small sample of the problem-solving undertaken by Scholander throughout his career. That such men as Irving and Scholander inspired the Schmidt-Nielsens and generously extended their invitation for the couple to join them in Swarthmore in 1946 proved significant for the future development of American animal physiology.

Once settled in Swarthmore, Knut Schmidt-Nielsen set right to work. In keeping with the lessons learned from Irving and Scholander on the importance of manageable equipment for fieldwork, he developed a portable kit for measuring chloride in body fluids as an indicator of salt and water regulation. But what to do with it? Irving told him: "It isn't enough that the chloride method works in the laboratory. We need to show that it works under field conditions. To test it, either we can go to the sea coast, where salt and water problems are obvious, or we can go to the desert, where there is little or no water at all." Irving seemed to have weighted the options in favour of the desert, so they travelled to Arizona in 1947.

In Arizona the Schmidt-Nielsens were intrigued by the kangaroo rat, which seemed to get by without water during drought periods. "Could desert rodents really live on dry food with nothing to drink," Knut asked, "and if so, by what unknown physiological mechanisms did they survive? This mystery begged for answers" (K. Schmidt-Nielsen 1998). They found indeed that the rats can survive with dry foods without drinking, that they do not store water in their body, and that they are able to limit water loss from the kidneys by excreting very concentrated urine (B. Schmidt-Nielsen et al. 1948). Add to this the nocturnal habit of the kangaroo rat and its escape into moist burrows during daytime, and you have an all-round desert animal. If this works well for a small animal, Knut reasoned, how do camels manage when their size denies them a place to escape from the heat? To research this question, the Schmidt-Nielsens – their family richer by a third child – moved for a year to Africa.

It took courage and determination, with three young children in tow, to find funds, make detailed trip preparations, and live for a year in the Algerian Sahara in the harshest of conditions. By that time (1953) Knut had taken posts at Stanford, the University of Cincinnati, and more recently at Duke University, where he remained for the rest of his career. Interestingly, when offered the position of professor of zoology at Duke, Knut asked that the title be changed to professor of physiology, even though the appointment was in the Department of Zoology. He justified this request by emphasizing that he was a physiologist, not a zoologist. To many colleagues this would have sounded odd, as to them being a comparative animal physiologist was part of being a zoologist.

Knut (1998) recalled how they envisaged their research program for the North African Sahara: "Our challenge was to determine whether camels rely on evaporation to keep the body temperature from rising unduly, similar to the way humans sweat to prevent overheating." Corollary questions popped up: Can camels afford to lose more water than other mammals? Do they store water in their body, such as in their hump or stomach? The answer to the latter question was negative. But the Schmidt-Nielsens (1956a) found that camels can tolerate "an extremely high degree of dessication of the body" in the desert heat, thanks in part to very low urine output and dry feces. "Particularly effective as a water conserving mechanism," they added, "is the low evaporative water loss during dehydration in the summer." If evaporation is not a problem, how can camels offset the heat load entering the body during the day, when the air temperature is, say, several degrees Celsius above normal body temperature? The camel's solution is to use their fur as a shield against heat gain and elevate their body temperature to 41°C, thus relieving the sweat glands of any call to excrete fluid to cool off (Schmidt-Nielsen et al. 1956b).

Knut was also curious to know how marine birds remote from sources of freshwater deal with salt load in the sea, either from drinking sea water or eating marine fish or invertebrates. Back in 1939 he had attempted to find out, but the answer escaped him. Now in 1956 he examined cormorants for answers. He found no evidence that cormorants drink seawater. However, they may ingest some sea water along with their food, and their invertebrate prey contain more concentrated salt than fishes. How do they deal with the excess salt intake that cannot be handled by the kidneys? Knut found that they "excrete a highly [concentrated] liquid that drips out from the internal nares and collects at the tip of the beak, from which the birds shake the drops with a sudden jerk of the head" (Schmidt-Nielsen et al. 1958). Similar salt glands were later discovered in the Galapagos marine iguana and a salt gland under the tongue in sea snakes.

At this juncture it is worth taking a moment to discuss an issue in the life of the Schmidt-Nielsens that had repercussions in the couple's life and resonates with historical lessons for husband-and-wife scientists. The salt gland paper of 1958 marks the beginning of Knut's research free of Bodil's involvement. As Knut relates candidly in his memoirs (1998), early in the 1950s "it quickly became apparent that my marriage was unraveling, a situation that adversely affected my work." Knut complained that Bodil increasingly withheld research data from him, and that she acted as a sole contributor. In addition, "Bodil often complained that because she was a woman, she was more likely to be viewed as a technician than as the full-fledged scientist that she was." When Knut sought psychiatric help, his treating doctor discerned that a rivalry had developed between Knut and Bodil, as both harboured an ambition to excel in the field they shared.

The resentment Bodil felt seems to echo that of a generation earlier, in the example of Knut's own mother. We have here an exceptional sample for examining such issues – an interconnected triangle of scientist couples. Both Knut's and Bodil's parents were trained as scientists. August and Marie Krogh, as it transpires in Bodil's biography of her parents, worked as a harmonious team on some research projects, while allowing for separate lines of inquiry. When August received his Nobel Prize, Marie expressed no recrimination that she should have had a share in it. Perhaps Bodil saw in her mother's acceptance of her place in her marriage and career – in itself advanced for that era – a negative model that coloured her assessment of her own marriage and professional quandary.

If the Kroghs experienced a good working relationship, Knut's parents' seemed by comparison stultified, and his mother simmered with muted resentment. His mother, after all, had worked with the great Arrhenius and felt entitled to use her intellect for greater things than just giving birth and raising children, however rewarding such domesticity can be. Knut recounted that she also "helped Father in his laboratory, where she served as a highly skilled technician." How eerily this echoes Bodil's fears of being perceived as a technician rather than a scientist. Knut added: "Mother felt that her collaboration with Father reflected only his interests, which concerned vitamins A and D. 'My real talents in math and physics were wasted,' she said." Her resentment was such that all the love she had for Knut's father had vanished.

Knut conceded in his memoirs that his mother's story, which she confided secretly to him in his late teenagehood, made such an impression on him that he was determined not to repeat his father's selfish stance with his own future wife. But complexities of marital context and of individual characters and issues can derail a well-intentioned plan such as Knut envisaged. Marriages between highly educated and bright people are always hazardous when one partner feels constantly hampered by social mores, but the risks ramp up when both parties work together and share the same ambition. Knut and Bodil's marriage never recovered; they separated in 1962.

Bodil's academic career took off with appointments as professor and eventually chair of the Biology Department at Case Western Reserve University in Cleveland, where she became recognized as a leader in the field of the comparative physiology of the kidney. Knut threw himself into his work, publishing his first book, *Animal Physiology* (1960), one of a series of slim biology textbooks designed for college classes. But the book that brought him the most fame was *How Animals Work* (1972), the summation of a series of lectures at Cambridge. Discussing its title in his memoirs, Knut gives the essence of the book's intent:

> The title I chose, *How Animals Work*, was a bit of a pun. Physiology is about how animal organs function or work, and I also discussed the work of running and flying and swimming. My lectures [at Cambridge University] had dealt with panting dogs and bird respiration, kidney function, whale flappers, the swim bladder of fish, animal locomotion, and a variety of other subjects.

The very first paragraph of the text encapsulates his approach to the investigation of animal functions:

A simple biological problem may arouse our interest, but as we gain more knowledge the questions ramify and appear to grow in complexity. This may take us to new and seemingly unrelated problems, but in retrospect they are all related to the desire to find out how things work. If we are fortunate we will gain some insight, and when we understand underlying principles, the greatest reward seems to be in the simplicity of the answers.

Knut used the problem-solving strategy of the engineer. Analogies with man-made devices such as heat exchangers abound in his depictions of physiological mechanisms. Unlike Prosser, he was not interested in the intellectual construction of a discipline. As he admits: "I am not made for difficult and complex problems. My interest in animals came to me naturally, and the questions I have tried to answer in my research have all seemed simple." The animals he was interested in were all upper vertebrates: mammals, birds, reptiles. He balked at the complexity that examination of a wider range of animals would inevitably bring. The word "comparative" rarely finds a place next to "animal physiology" in his texts.

How Animals Work was a great success as a semi-popular book. Between 1973 and 1993 it was reprinted ten times. Its accessibility to a wider readership owed much to the clarity and liveliness of Knut's writing style and to the fact that it dealt with animals to which readers could relate more easily than fishes and invertebrates. In 1975 he published a more comprehensive textbook – *Animal Physiology: Adaptation and Environment* – which served as a counterweight to Prosser's unwieldy textbook, which had been geared more toward research students and scholars than to the undergraduate class. "Not only did it enjoy widespread use," Steven Vogel (2008) wrote, "but several successful alternatives taking its approach have appeared subsequently. In a sense, it provided a capstone for the amalgamation of physiology and zoology begun long ago by Bohr and Krogh."

∼

A contemporary of Knut Schmidt-Nielsen who had less visibility beyond the confines of academia, but whose contribution to the field is no less significant, was George A. Bartholomew (1919–2006). Their paths to environmental physiology converged, but they started from quite distinct sources. Bartholomew was born in Missouri and his family eventually moved to Berkeley, California (Dawson 2011). He completed his undergraduate studies at the University of California at Berkeley in 1940 and had barely started a PhD at Harvard when it was interrupted by World War II. After finally earning his PhD in 1947, he joined the faculty at UCLA's Zoology Department, where he remained for his entire career.

Bartholomew's first publication, accomplished while he was doing his master's at Berkeley, reflected his zoological/ecological interests; it described how cormorants flock, circle, and dive for food in San Francisco Bay (Bartholomew 1942). Once at UCLA, he investigated Schmidt-Nielsen's pet animal, the kangaroo rat, but from a totally different perspective, namely the evolutionary and ecological implications of the rodent's locomotion (Bartholomew and Caswell 1951). His first publication of a physiological nature – how juvenile marine birds struggle to maintain a stable body temperature in arid environments – appeared in 1954. In a seminal essay paper (1958) Bartholomew argued that physiological tolerance plays a determinant role in the distribution of animals in ecosystems only in aquatic animals and terrestrial invertebrates, whereas the distribution of terrestrial vertebrates is best explained by ecological and behavioural factors. He went on over the years to work on temperature regulation and energy metabolism in hosts of birds, reptiles like the Galapagos marine iguana, and mammals like the pinnipeds (sea lion, elephant, and other seals), always with an eye to their social behaviour.

The above makes clear how the viewpoints of Schmidt-Nielsen and Bartholomew differ. As historian of biology Joel Hagen (2015) points out: "The differences between Schmidt-Nielsen's engineering approach and Bartholomew's explicitly evolutionary and ecological approach reflected broader disciplinary divides in organismal biology." What are these divides? Hagen explains that, as an example, Schmidt's engineering mindset channelled him to look for simplicity, for "elegant designs" in animal functions, whereas Bartholomew, zoologist to the core, embraced the messy complexity of animal lives that were subjected to a host of ecological constraints and evolutionary forces. Bartholomew was the first to stress the role of behaviour in functional regulation – shade-seeking, for instance – noting that, in contrast to physiological adaptations, which develop slowly, "behavioural adjustments to the environment can be drastic, rapid, precise, and of exquisite flexibility" (Bartholomew 1964).

This divide was reflected in institutional settings as well. As Schmidt-Nielsen was quick to project himself first and foremost as a physiologist, he gravitated around the American Physiological Society and published many of his papers in the society's organ, the *American Journal of Physiology*. The society catered primarily to human or paramedical physiology, but Schmidt-Nielsen organized an informal Comparative Physiology Group as soon as 1950. Only in 1977 was an official Comparative Physiology section implemented by the society and incorporated into the make-up of the society's journal (Cook 1987). Bartholomew, in contrast, was affiliated with the American Society of Zoologists and published in non-physiological journals such as *The Condor* and *The Auk* (devoted to birds), the *Journal of Mammalogy*, and the *Journal of Experimental Zoology*.

In spite of differences of outlook harboured by these pioneering founders of the American Schools of Animal Physiology – Prosser, Irving, Scholander, Schmidt-Nielsen, Bartholomew – they and their followers managed to form a cohesive community of a sort. Does that mean that the field of comparative animal physiology was ready to be elevated to the status of a scientific discipline? Not quite, if one adheres to the stringent criteria enunciated by historian of science Timothy Lenoir. Lenoir (1997) explains where disciplines stand in the grand operational scheme of science:

> Within this complex of issues generated by the disunity of science, discipline emerges as a crucial site; for just as laboratories and sites of apprenticeship are essential for organizing and reinforcing the economies of skill necessary for conducting science locally, disciplines are the structures in which these skills are assembled, intertwined with other diverse elements, and reproduced as coherent ensembles suitable for the conduct of stable scientific practice more globally. Disciplines are the infrastructure of science embodied above all in university departments, professional societies, textbooks and lab manuals.

By 1960 comparative animal physiology still lacked sufficient coherence globally. Neither university departments wholly dedicated to the field nor independent professional societies had yet emerged. Chairs of comparative physiology existed, but they were embedded in departments of zoology or, less often, in medical schools. On a more positive note, curricular niches, textbooks, and laboratory manuals had made an appearance. And even more significantly, as a field, animal physiology was thriving in the United States. How did it fare elsewhere?

6

The Belgian School and the Rise
of Comparative Biochemistry

∾

It is only a matter of applying to invertebrates the various and ingenious methods
created by modern vertebrate physiology.

Léon Fredericq (1878)

Oddly, a small European country – Belgium – played a disproportionately
large role not only in the development of comparative animal physiology
but also in seizing the leadership of an offshoot, the rising field of comparative biochemistry. In fact, this locus of activity must be further circum-
scribed to the city of Liège. One may ask what exceptional qualities made
Liège such a hotbed for the field. Since the eleventh century (Geenen 2015),
Liège had established many schools and libraries that attracted students and
scholars from all over Europe. Indeed, Petrarch, the famous medieval Italian
poet, was so impressed by the intellectual life of the city that he called Liège
the "Athens of the North." And, Vincent Geenen suggests, once the University
of Liège was established in 1817, it flourished thanks to the city's "tolerance
for the different philosophical and political movements which is truly its
'identity card.'" In this atmosphere, which can only be considered conducive
to solid scholarship, physiology was part of the curriculum from the start
(Florkin 1979).

The first luminary of biology to make his mark in Liège was Theodor
Schwann (1810–1882). From his microscopic studies in Berlin under Johannes
Müller between 1834 and 1839, Schwann reached the conclusion that all an-

imal and plant tissues are formed from cells containing a nucleus (Schwann 1839). Through this work, Schwann is credited, along with botanist Matthias Jacob Schleiden (1804–1881), for having formulated the cell theory, one of the great milestones in the history of biology. A fervent catholic, Schwann found it difficult to live exposed to the staunch Protestantism of the Prussian capital, so he moved to Belgium, first to Louvain and then, in 1848, to Liège. There, he held the chair of physiology, general anatomy and embryology for thirty years. Upon his retirement, he was so impressed by the research work of a young Belgian physiologist that he designated him as his successor (Nolf 1937). That young Belgian, Léon Fredericq (1851–1935) went on to be the founder of the Belgian school of comparative physiology.

Fredericq's prolific accomplishments in mammalian as well as comparative animal physiology can be said to rival those of the Dane August Krogh (discussed in chapter 4). Born in Ghent to a physician father of French origin and a Flemish mother (Nolf 1937), Fredericq (not to be confused with his son Henri, of whom more below) followed the pattern of many gifted scientists who showed a predilection for maths and natural sciences in school and an addiction to collecting shells, insects, plants, and minerals throughout their childhood. Zoology and chemistry were his favourite topics at the University of Ghent, where he earned his doctorate in natural sciences at the age of only twenty and his medical degree at twenty-five. Although he was at first attracted to comparative anatomy and cataloguing animals, his medical studies redirected his interests toward the "new medical science," physiology (Nolf 1937). But an obstacle stood in his way: his revulsion at the sight of blood was such that it even led to fainting. He conquered this fear, however – and, ironically, blood became a major focus of study for him.

In 1876 Fredericq was granted a research fellowship that allowed him to train under renowned professors abroad. He went first to Strasbourg and the laboratory of Felix Hoppe-Seyler, who was preparing the publication of his treatise on physiological chemistry (1877–81). Physiological chemistry – the precursor of biochemistry as a discipline – was thriving in Germany way ahead of other European and Anglo-Saxon countries (Kohler 1982). Originally cast in a supporting role in physiological institutes, physiological chemistry was trying to break free and Hoppe-Seyler led the charge by founding his own Institute of Physiological Chemistry in Strasbourg. "In 1877," Kohler

Léon Fredericq.
Wellcome Library 12757i.

wrote, "Felix Hoppe-Seyler singled out the possessive attitude of physiologists as the main reason why more institutes of physiological chemistry were not being created." Perhaps Fredericq retained lessons from this turf war, as comparative physiology and biochemistry coexisted seamlessly in his and future generations in Liège.

Fredericq's second stop was Paris, where he learned the experimental methods of physiology from the laboratories of Paul Bert (discussed in chapter 3) and Etienne-Jules Marey (1830–1904) at the Collège de France. In Paris he was exposed not only to classical mammalian physiology but also to the comparative physiological studies of his hosts. In particular, Bert's interest in cephalopods and the respiratory function of blood would be echoed in Fredericq's own research. But his encounter with Marey must have been especially entertaining, as Marey was probably the most original and eccentric physiologist he ever met. Marey represented a fresh, French, vision of the biomechanical ways of humans and animals, and for this reason alone a detour into his life and accomplishments is warranted.

Marey never lost an opportunity to think outside the box. Burgundy-born, he wanted to train as an engineer but his father brought his formidable will to bear in the direction of medicine (Laporte 1998). As a hospital internist in Paris, at only twenty-nine, he invented a pulse recorder (sphygmograph), which allowed the pulse rate to be recorded on smoked paper in real time. This invention came only twelve years after Carl Ludwig's kymograph (chapter 3). By patenting his device, Marey earned sufficient royalties to set up a private laboratory and experiment to his heart's content with no financial worries. In his book *La méthode graphique* (1885) he gave an exhaustive exposé of the methods available at the time for plotting data on a chart and recording countless movements continuously over time – blood pressure or flow, air movement during respiratory cycles, muscle contraction/relaxation and force, to mention a few. His long catalogue of graphic methods stemmed, he claimed, from a mistrust of the ability of the human senses to observe natural phenomena. Fredericq's exposure to such expertise later proved invaluable for conducting his own physiological experiments.

In another remarkable book, *La machine animale* (1878), Marey revisited the territory covered by his countryman Julien Offray de La Mettrie in *L'homme machine* 130 years earlier (see chapter 1). But Marey's approach was

Etienne-Jules Marey in 1878.
Reproduced from picture preceding part I in Braun (1992).

more reminiscent of Giovanni Borelli's pioneering biomechanics from the seventeenth century than of La Mettrie's theoretical construct. Marey, with the array of methods at his disposal (borrowed or of his own creation), set the stage for the modern field of biomechanics. At the time of Fredericq's sojourn, Marey was at the height of his reputation as an investigator of animal locomotion. In *La machine animale* he stressed the importance of learning from nature when designing human-made apparatus: "If the hull of a ship is, as is rightly said, shaped on the model of a swimming bird, if the invention of sails and of oars mimics the wing of the swan inflated by the wind and its membranous leg thrashing through water, it is but a small part of the crafts borrowed from nature" (Marey 1878).

In his own evocative words Marey illustrated the diversity of aquatic locomotion in the animal world:

Here is a fish who strikes water with the flat surface of the tail, there is an octopus, a cuttlefish, a jellyfish who, briskly squeezing a pouch full of fluids, push the water in one direction and are thereby propelled in the opposite direction; the same happens when a mollusk quickly closes the valves of its shell and is propelled in the direction opposite the water current it created. Dragonfly larvae eject from their intestine a liquid jet of such force that they are propelled forward over a good distance.

Whichever mode of locomotion is analysed, Marey reminds us, forces are exerted and the energy costs entailed have to be factored into the balance sheet. The extreme mobility of water and air, for instance, makes conditions for swimming and flight less favourable than conditions for locomotion on land. The resistance met in air or water and the fulcrum or pivot for motion thrust can vary widely:

[F]ish possessing roughly the same density as water find themselves suspended in it without having to spend energy to prevent sinking and if they need to move in a certain direction, they only need for this purpose to overcome the resistance of the fluid to be displaced. Birds, in contrast, just to stay aloft, must generate sufficient work to constantly counteract gravity. If in addition they are in motion, birds must spend the extra energy needed to overcome air resistance.

By means of a pressure sensor inserted in a shoe sole pad of his invention, Marey was able to measure in real time motor force in the legs during biped walking and running, and foot pressure on the ground. He also analysed the quadruped trot and the gallop of horses, with multiple sensors recording every component function in the rhythmic movements. Sensor devices decorated the legs and the horseshoes, while a mounted rider carried the recorder wired to the leg sensors. Factor in the instruments measuring the muscle response of the withers and croup to the horse's motion cycle, and Marey had a well-decked-out animal.

Marey also asked intriguing questions about insect flight: What is the frequency of wing movements? What is the series of relative positions of wings in space through the wing beat cycle? How is the motor force that supports and moves the insect body generated? He measured wing beat rate by tethering an insect in such a way that, as the wing beats, it touches a lightly smoked paper on a revolving drum, thus leaving a white trace with every beat. He found that the frequency ranged from nine per second in the butterfly to a whopping 330 per second in the common fly. He demonstrated how the wing bends differently between up and down movements of the beating cycle. "To sum up," he concluded, "a pendulum oscillation by the wing veins in tandem with air resistance suffices to generate all movements observed in flight."

To confirm his findings Marey designed an artificial insect. Here is his laconic explanation of the artifact: "Let's suppose an apparatus which, under the control of a connecting rod and crank, induces in a stem rapid, flexible up and down movements. Let's affix a membrane similar to an insect wing to this stem which now acts as the vein; we will see reproduced all the movements that the insect wing executes in space." Marey showed that it is air resistance that curves the wing in a double ellipse figure (figure-eight sequence). He adopted a similar biomechanical approach to study the flight of birds loaded with his uncanny sensors.

What particularly caught people's imagination, however, was Marey's ingenious use of photographic technique, such as it was in the 1880s, to further probe how animals move (Braun 1992). According to historical archives of the Naples Zoological Station researched by Marina Vagnoni, this fascination began when Marey became enamoured with Madame Joseph Vilbort, the wife of a novelist and newspaper editor. The story goes that he followed her to the Gulf of Naples, considered the best climate to alleviate her diagnosed neurological disease. (Her husband symmetrically fell in love with another woman, whom he later married.) Madame Vilbort and Marey maintained their relationship for twenty-five years, until her death. From 1880 they lived in a house in Posillipo purchased by Marey.

In Naples Marey struck up a friendship with the director of the Zoological Station, Anton Dohrn, from whom he obtained animals for his research. One of the animals in the station's aquarium was a ray whose peculiar way of swimming he analysed. But what earned him the badge of the eccentric scientist in Naples was his invention of the chronophotographic rifle to visually record the movement of birds and their wings in flight. The idea was to obtain a sequence of photographs, a fraction of a second apart, of a bird in the air. Marey met that challenge in 1882. He constructed a rifle in which the cartridges were replaced by a lens, to which a rotating twelve-hole disc containing a film was attached. A pull on the trigger unleashed a rapid rotation of the photographic disc, so that twelve successive images of the bird were obtained in one second on a single film.

Here is how Marey described his invention to his mother: "I have a photographic gun [*fusil photographique*] that has nothing murderous about it and that takes a picture of a flying bird or a running animal in less than 1/500 of a second. I don't know if you can picture such speed, but it is something astonishing" (quoted by Braun 1992). The achievement was made even more amazing for the time by the development of a film plate capable of keeping up with the speed of the camera.

An anecdote that made the rounds in Naples tells of Marey seen "shooting" seagulls. Observers watched as he aimed his rifle, but they neither heard shots nor saw smoke coming out of the barrel, and no bird fell; yet Marey seemed quite content with himself. As a result he was nicknamed the madman of Posillipo. But the "madman" was not intimidated by puzzled onlookers, and his mysterious rifle went on to photograph the movements of a host of other animals. As media historian Marta Braun has shown, this feat led to Marey's development of sequential photography revved up to sixty frames per second, thus producing smooth enough animation to earn Marey a place among the precursors of the cinema (Braun 1992). Marey himself would have felt awkward being considered a precursor of cinematography, as he was interested in photography only as another technique – admittedly enriched by his amazing creativity – in his physiological toolkit.

~

Marey's influence on Léon Fredericq made itself felt in the importance that the latter attached to graphs and recording techniques in his own research. He explicitly stressed this point in his manual of laboratory experiments in physiology destined for students (Fredericq 1892), in which he gave Marey due credit. And both Paul Bert and Marey influenced his career choice. They inspired him to devote his life to basic research to such an extent that, when he was offered a medical appointment on his return to Ghent, he declined (Nolf 1937). Research work was fulfilling for him even though he had to be content with a smaller salary. As he commented later in life: "[B]y devoting myself to science it never occurred to me that I was sacrificing something meritoriously. What merit is there to follow one's tastes, to surrender to one's own fantasy? In our profession, [Walter] Cannon said, there is something miraculous; we are paid to do precisely what pleases us the most" (quoted in Florkin 1979).

Fredericq first intended to conduct human or mammalian physiological research, but a visit to the Zoological Laboratory of Roscoff in Brittany in 1878 convinced him of the opportunities for discovery that marine fauna opened to the physiologist. Throughout his career he followed the Krogh model of alternating his research between mammalian and comparative or (zoo)physiology. But early on, he showed his mettle by contributing substantial new knowledge on mechanisms of blood coagulation as well as studies of protein composition and dissolved gases in the plasma and red blood cells. These contributions formed the basis of his doctoral dissertation in physiology, presented at the University of Ghent in 1878. Theodor Schwann was so impressed by Fredericq's thesis that he invited the young man – who was only twenty-seven – to succeed him at his chair in Liège. When the famous old man retired, Fredericq occupied the chair in 1879 and he remained at the Faculty of Medicine of the University of Liège until his own compulsory retirement in 1921 (Nolf 1937).

Before taking the position in Liège, Fredericq had already made the best of his time in Roscoff. The "Laboratoire de zoologie expérimentale," founded by the distinguished zoologist Henri de Lacaze-Duthiers (discussed in chapter 3) only a few years before Fredericq's visit, was Lacaze-Duthiers's instrument for restoring the prestige of French biology after France's defeat to Prussia in 1870 and for putting to rest Claude Bernard's assertion that zoology is a descriptive, not an experimental science. In this context, Fredericq the physiologist was certainly welcome, and from his description of the local facilities it can be surmised that the staff went out of their way to make his time at the station profitable:

> The research that I publish on the physiology of Octopus was conducted at Roscoff during July and August 1878. Professor Lacaze-Duthiers, who was kind enough to grant me for the second time some space in his laboratory of experimental zoology, put at my disposal all the resources at hand: dissection instruments, physiological equipment, aquaria, fishing gear ... Lodged on the premises of the station, with all my experimental animals within visual range, with the freedom to consult books in the library at all hours of day and night, I was blessed with exceptionally favourable circumstances for my studies. (Fredericq 1878)

And he delivered handsomely. Primary was his discovery of the equivalent of haemoglobin in the blood of the octopus. He called the oxygen transporter in the octopus haemocyanin, to stress that the protein gives the invertebrate's blood a bluish colour – instead of red with haemoglobin – when oxygenated. The bluish tone, he found, is due to the copper that haemocyanin contains, as a substitute for the iron of haemoglobin. This was the first of many discoveries showing how diverse such oxygen carriers are in the blood of different kinds of invertebrates.

Another interesting discovery that Fredericq made as early as 1882 was autotomy (Fredericq 1883), the ability of certain animals to let go of some part of their anatomy that is under attack by a predator. The strategy for this seemingly odd self-mutilation is a no-brainer: it is better to sever a body part already in the jaws of the attacker – the tail in some reptiles or a leg in crabs, for example – than to be chewed whole. Not only did Léon describe the phenomenon but he also tracked the mechanism in crabs: it involves an automatic nerve reflex that activates a muscle powerful enough to squeeze out the body part. In time the body part regenerates.

Yet another of Fredericq's major contributions concerns how the blood and tissues of aquatic animals cope with surrounding waters of varying salinity. The question preoccupied him from his first trips to the Naples Zoological Station in the early 1880s. He came up with a three-part evolutionary scheme: (stage A) marine invertebrates show no salt regulation by conforming their internal salt concentration to that of sea water; (stage B) sharks, whose blood is less saline than sea water, maintain the same osmotic concentration as sea water by replacing the missing salts with urea and other organic solutes; and finally (stage C) bony fishes, both marine and freshwater, maintain the osmotic and salt concentration of their blood at one third of that of sea water (Fredericq 1901). Fredericq was the first to point to gills as regulators of water and salt movements in addition to gas exchange (Fredericq 1891).

The German invasion of Belgium in 1914 dealt a terrible blow to Fredericq's research career, which he was unable to resume until 1920. Even then, it was but a pale reflection of his earlier output. After his retirement in 1921, his son Henri succeeded him as chair of physiology in Liège, thus ensuring the continuity of the physiological tradition established by his father. But Henri Fredericq (1887–1980) was no match for his formidable father. It did not help that he started his career as a teaching assistant (1908–12) and a research assistant (1912–19) in his father's shadow (Bacq 1983). In an effort to prove himself capable of independent scholarship, he took a position as assistant professor of physiology in Ghent. Two years later he was called back to Liège to take his father's chair.

Henri was clear-eyed about his research accomplishments relative to his father's. In a letter dated 13 December 1928 to the president of the Belgian Royal Academy, he wrote: "Of course I am well aware that I have devoted to the Science I loved and cultivated over many years my very best. But it is also true that the progress I may have imparted to human and comparative physiology is not particularly earth-shattering." Henri's range of physiological interests, confined largely to the physiology of the nervous system, was narrower than his father's. And a perusal of his bibliography reveals that for him paramedical physiology predominated over comparative animal physiology.

The mainstay of Henri's research was cardiovascular physiology, especially the modulation of heart function by its innervation. His primary animal models for the basic studies were the turtle first, followed by the dog. These studies led in time to neuro-pharmacological studies and the mechanisms of neurotransmission. He was, for example, the first to show the effect of caffeine on neurotransmission in the heart (Fredericq 1913). When he conducted research at the Marine Biological Laboratory in Woods Hole, Massachusetts (Fredericq 1930), and later in Naples, he widened his cardiovascular studies to include invertebrates. It was his discovery that the heart of the horseshoe crab, the octopus, and several other invertebrates has a double innervation: one nerve accelerates the heart and another slows it down (Frederick 1947).

Henri's main legacy, however, was to have ensured that the Liège School of Physiology was perpetuated. From his early years in the chair of physiology, he was subject to the vagaries of teaching assignments that severely constrained him. At the Faculty of Medicine he taught general and mammalian physiology and physiological chemistry, and at the Faculty of Science animal physiology and "biological chemistry" (Bacq 1983). This excessive teaching load resulted in a perverse way from his great popularity as a teacher. In the 1930s, when Henri was able to persuade the administrators of the need to lighten his load, he steered them toward successors who he thought showed great promise. In 1934 his pupil Marcel Florkin (1900–1979) was appointed chair of biochemistry at both the Faculty of Medicine and the Faculty of Science (Zoology) to allow Henri to divest himself of the "physiological chemistry" that he never felt competent enough to teach. And in 1939 Zénon M. Bacq (1903–1983) – at the time a collaborator of Henri – was given the new chair of animal physiology at the Faculty of Science. As it turned out, Henri's gamble paid rich dividends.

∾

Florkin exhibited greater similarity to Léon than to Henri Fredericq. Like the former, he had a prodigious capacity for work, he breathed authority, and he had an unmistakable entrepreneurial spirit. In addition, some of Léon's research interests, such as oxygen carriers and other blood proteins in invertebrates, were considered to be within the purview of the "chemical zoology" that became an important focus of Florkin's research activities.

Comparative biochemistry, which absorbed chemical zoology, flourished under Florkin's care but he was not the founder of the field. Before appreciating the consolidation of the field of comparative biochemistry under Florkin, however, it will be worth our while backtracking to its origins.

The earlier success of physiological chemistry as a discipline owed much to the stewardship of its late-nineteenth-century German protagonists, especially that of Felix Hoppe-Seyler in Strassburg (Kohler 1982). Unlike many of his colleagues, Hoppe-Seyler was trained in biology as well as chemistry. Although the word *Biochemie* (biochemistry) first surfaced in the inaugural issue of Hoppe-Seyler's *Zeitschrift für physiologische Chemie* (1877), no one else used the word in any publication for another quarter of a century – not until Hoppe-Seyler's successor in Strassburg, Franz Hofmeister, incorporated it in the title of his own periodical (*Zeitschrift für die gesammte Biochemie*, 1902).

It was also in 1902 that Otto von Fürth (1867–1938), Hofmeister's assistant and lecturer, put the final touches to the first book ever devoted to comparative biochemistry. Bohemia-born, Fürth had studied medicine and chemistry at the University of Vienna, writing his thesis under Hofmeister's supervision (Adler-Kastner 2000). Fürth followed his mentor to Strassburg to assist him and serve as lecturer. At the turn of the century he was among the first to extract and semi-purify the hormone and neurotransmitter adrenaline, which he called suprarenin. In the preface to his seminal 1903 book *Vergleichende chemische Physiologie der niederen Tiere*, Fürst explained how a medical chemist like himself came to compose such a huge opus:

The long-cherished plan of attempting to compose a "Comparative Chemical Physiology of Lower Animals" took firm shape in me when, during a visit to the Zoological Station in Naples, I became personally acquainted with a number of notable biologists who impressed on me that the need for a physiological interpretation of the problems of modern zoology had long been due. So, urged in the friendliest way by Privy Councillor Dohrn in Naples and Professor Hofmeister, I harnessed myself to this task.

With heroic German thoroughness, Fürth summoned all the literature on the topic available at the time and stuffed it between the covers of the book with barely a modicum of critical evaluation. Every researcher who had ever tackled invertebrate organs – respiratory, digestive and excretory organs, blood, muscle, or gonads – in order to analyse their chemical composition was represented, including himself. As an informative compendium, the text served the scientific community well, but little in the way of functional significance stood out from the mass of information. The book's last chapter merely alludes to the conditions for life and survival from the biochemical point of view; that is, to the essential inorganic and organic substances, and how they are managed by the animal machinery in changing environmental conditions.

The comparative biochemist in Fürth had shallow roots; in 1905 he returned to Vienna "to become head of the Chemistry Section of the Physiology Institute, attracting many research students from Europe and overseas" (Adler-Kastner 2000). He enjoyed a high-profile and productive career in medical biochemistry, but in 1938, six months short of his retirement age, he suffered the fate of half the senior faculty of the Vienna Medical School: as a Jew, he was sacked in the wake of the Anschluss. The blow must have been devastating because within months he died from a stroke.

For lack of a conceptual grounding, comparative biochemistry was off to a poor start. What physiological chemists such as Fürth had accomplished until then was to extend to "lower animals" the dry survey of chemical processes discovered in mammals and humans, with no thought to whether these processes contributed to the understanding of how animals work. For the conceptual ground to clear the way for the embryonic field, it seems that the field had to move to the UK. The thread leading back to the intellectual source of comparative biochemistry can be followed upstream to Michael Foster, the founder of the chair of physiology at Cambridge University. Foster's unique vision of physiology as a science comprehensive enough to include physiological chemistry opened up the field to a promising new generation:

Believing that physiological processes would ultimately be explained in chemical terms, Foster sent three of his favorite students, Walter Gaskell, John Langley, and Arthur Sheridan Lea, to learn chemical physiology at Kuhne's knee at Heidelberg. He nudged other protégés into chemical problems. When Walter Fletcher decided to take up muscle work in

1898, he asked Foster if he thought there was any promise in the chemical side; Foster, he later recalled, "rolled up his beard with both hands
over his mouth and chuckled." The ultimate result of Foster's eloquent
silence was the collaboration of Fletcher and F.G. Hopkins on the biochemistry of muscle contraction. (Kohler 1982)

Thanks to Foster's incubation of the field, Frederick G. Hopkins (1861–
1947) became the founding father of British biochemistry and in 1914 occupied the first chair of biochemistry at Cambridge (Dixon 1997). Hopkins was
celebrated for his discovery of vitamins. He saw clearly where biochemistry
stood in relation to physiology: "Physiology as ordinarily understood is
chiefly concerned in every case with the visible functioning of organs; biochemistry rather with the molecular events which are associated with these
visible activities" (foreword to Baldwin 1937). Hopkins's student Joseph
Needham (1900–1994), in turn, facilitated the establishment of comparative
biochemistry in England. Heralded in his time as the new Erasmus, Needham is better known to a wide readership as the hero of Simon Winchester's
The Man Who Loved China (2009). However, we will dwell here briefly, not
on Needham's expertise on Chinese science and civilization or his entertaining eccentricities with which Winchester peppers his book, but rather on his
early scientific achievements.

Needham was primarily interested in the chemical events associated with
embryology and growth, and his wife, Dorothy Moyle Needham (1896–1987),
in the chemical events occurring in muscle cells during contraction. These
were milestones in their own right. The comparative biochemical approach
taken by Needham owed more to his interest in evolutionary theory than to
any craving for functional explanations. He was the first in a long line of
biochemists, including Florkin in particular, for whom evolutionary considerations constituted the abiding motive for their comparative outlook.

One of these evolutionary considerations was the controversial theory
of recapitulation, according to which the developing embryo recapitulates
ancestral traits before maturing as an adult into the version of the trait fitting
for its level of phylogenetic organization. Using popular imagery, Needham
(1930) likened it to "an animated cinema-show of ancestral portraits." But
he found the concept inadequate, as it has little explanatory value and, after
all, embryos and larvae possess unique morphological traits that to an extent

Joseph Needham in Cambridge, 1965.
Internet source: Kognos.

define them as having animal machineries that function differently from those of adults. Apart from the fact that this distinct identity of the embryo justified Needham's curiosity about how embryos work, the embryo's idiosyncrasies tended to distort the procession of ancestors in the "portrait gallery." He realized that: "The chick embryo does not, then, pass through a fish stage, but through a stage resembling the embryonic form of a fish, and it is consequently at the cost of biological accuracy that writers such as [novelist] Aldous Huxley, wishing to emphasise the contrast between man's activities and his origins, refer to him effectively enough as an 'ex-fish.'"

Needham tackled the problem by intuiting that biochemical traits are better candidates to clarify the causes of recapitulation. He looked no further than bird eggs, the mainstay of his research program. Examining modes of nitrogen excretion by the kidney, he found that as bird embryos develop they go through a time series of excretory products: first, ammonia, the adult mode in aquatic invertebrates and some fishes; then urea, as found in adult sharks, frogs, some reptiles, and mammals; and finally, uric acid, carried over to adult birds. It is a sequence of increasing complexity in terms of enzymatic steps and energy expenditure to produce these nitrogenous compounds. Needham offered his explanation on how recapitulation works:

> The reason why the chick does not excrete uric acid from the very beginning would, therefore, be that it has not until a certain point developed the machinery for doing so, not that a urea stage was essential "physiogenetically" ... Recapitulation may be regarded as fundamentally the result of the necessary passage from simplicity to complexity, from low to high organisation, which is entailed by the animal's sexual system of reproduction, with its single egg cell.

But why bother with uric acid as an excretory product if it is such a complex process to produce it? The French physiologist and chemical zoologist Henri Delaunay found out. He noted that uric acid dominates in terrestrial animals that need to save water, such as birds, insects, and other animals in arid environments (Delaunay 1931). He deduced that uric acid was worth the investment because it takes very little water to shed it through the kidneys, unlike ammonia and urea. This is the kind of biochemical puzzle that induced Joseph and Dorothy Needham to ask if some biochemical markers could shed light on the origin of vertebrates (Needham and Needham 1932). After reviewing the various scenarios of potential invertebrate ancestors put forward over the years, the Needhams decided to use phosphagens as markers for tracking vertebrate ancestors in the genealogical tree and settle the issue.

Phosphagens are high-energy phosphate compounds stored in muscle cells which provide the phosphate needed to fuel the energy metabolism of active muscle. Invertebrates use arginine phosphate as phosphagen, and vertebrates use creatine phosphate. The Needhams travelled to the Biological Station of Roscoff on the coast of Brittany to survey as wide an array of species as possible (Needham et al. 1932). They found only two animal types in which both phosphagens coexist in muscles: echinoderms (sea urchins) and the strange-looking hemichordates (acorn worms). The cephalochrodates (Amphioxus), which are distant relatives of hemichordates, share with all vertebrates the exclusive presence of creatine phosphate in their muscles. Thus the Needhams came to the conclusion – later proved to be premature (Barrington 1975) – that vertebrates arose from a lineage of echinoderm and protochordate ancestors.

One of the co-authors of this scientific study of the Needhams' was Ernest Baldwin (1909–1969). Baldwin did both his undergraduate and graduate studies at Cambridge University (Kerkut 1970). For his doctoral thesis, supervised by Frederick Hopkins at the Institute of Biochemistry, he chose the comparative study of phosphagens (Baldwin 1933) and this was how he came to work with the Needhams. At the institute he was clearly overshadowed by the brilliant couple; his scientific output was modest in comparison. As a result his academic career picked up slowly; only on reaching forty-one did Baldwin get a chair of biochemistry at University College in London. But what he lacked in research excellence he more than made up for with his gift for communication. He excelled in class teaching and scientific publishing. When still a low-ranking lab demonstrator at Cambridge, he published *An Introduction to Comparative Biochemistry* (1937).

Baldwin's book, considered a classic in the field, has gone through reprints and three editions. Its enduring success is due to its simplicity; it avoids Fürth's pitfall of producing a heavy tome of amassed data that lacks perspective. Baldwin reached his audience by selecting a few themes – the role of ions and water in the colonization of freshwater and land; the excretion of

nitrogen; oxygen and carbon dioxide transport; animal pigmentation; nutrition; digestion; and metabolism – and showing their functional significance to the animal. In the book's introduction he was explicit about his
credo and the challenge the practitioner faced:

> [T]he task of the biochemist is, after all, the study of the physicochem-
> ical processes associated with the manifestation of what we call life –
> not the life of some particular animal or group of animals, but life in
> its more general sense. From this point of view a starfish or an earth-
> worm, neither of which has any clinical or economic importance per
> se, is as important as any other living organism and fully entitled to the
> same consideration, and unless such forms do receive considerably
> more attention than is accorded to them at present, biochemistry, as
> yet hardly out of its cradle, will assuredly develop into a monster.

The mission of comparative biochemistry, Baldwin asserted, was twofold:
to study "the reciprocal relationships between organism and environment
from a physicochemical standpoint" and "the physicochemical approach to
evolutionary problems." The latter, he insisted, was "peculiarly the business
of comparative biochemistry," adding that "probably few notions have had
such wide repercussions upon the world of human thought as the theory of
evolution." Historically, *An Introduction to Comparative Biochemistry* stands
as the theoretical foundation of the field. Baldwin went on to publish another
classic text, *Dynamic Aspects of Biochemistry* (1947). In 1969, after years of
suffering the effects of myotonic muscular dystrophy, he died, his career cut
short by congestive heart failure. Such was the British legacy handed down
to Marcel Florkin in Liège.

~

The following narrative of Marcel Florkin's life and legacy draws liberally
from two key documents: a memoir of the Belgian Royal Academy (Bacq
and Brachet 1981) and Florkin's own short autobiography (Florkin 1973).
Florkin was born with the twentieth century in the very city where he flourished and where his father was a city clerk, Liège. He felt the parental pres-

sures for a conventional profession, and his experience rings true for many of the protagonists in this book: "My father hoped to see me become an engineer, the most glorified profession in an industrial environment, or at least a doctor, and it was to please him that I went through the medical curriculum, as being nearest to my biological preferences."

Physiological research that he conducted in Henri Fredericq's lab during his medical studies resulted in his first publication. After he graduated, he obtained a research fellowship to study in the United States, where Henri had arranged for him to work in the lab of Edwin Cohn at the Department of Physiology of Harvard Medical School. There in 1928–29 he worked on protein chemistry and developed a relationship with the chair of the department, the famous Walter B. Cannon, and his wife, writer and social reformer Cornelia James Cannon, whose life was recently celebrated (Diedrich 2010). Another fellow with whom Florkin developed a deep friendship at Harvard was Alfred C. Redfield (1890–1983). Florkin owed much to Redfield for the trajectory of his career.

Redfield, born in Philadelphia, was a pure Harvard product, having earned his bachelor's degree and his PhD at Harvard's Museum of Comparative Zoology. In the course of his thesis research, Redfield discovered that skin colour change in amphibians (toads) is controlled by hormones (Redfield 1918). In 1921 he moved to the Physiology Department, where he was hired by Cannon as an assistant professor (Revelle 1995). After a few years of medical physiology, inspired by Léon Fredericq's work, Redfield embarked on a series of studies on the biochemical properties of haemocyanins, using the Marine Biological Laboratory in Woods Hole, where his species of interest, the horseshoe crab, could be found in abundance. It was this new research interest that brought Redfield and Florkin together. As Florkin (1973) recalled:

Alfred Redfield, with whom I had become acquainted in Cannon's department, [invited] me to go with him, for the summer months of 1930, as research worker, to the Hopkins Marine Station in Pacific Grove, near Monterey (Cal.) … we decided to travel together by car, a trip which lasted a whole month … Alfred Redfield had accomplished classical work on hemocyanin. His fine personality and his broad biological outlook delighted me. In the Hopkins Marine Station, Redfield had

Alfred C. Redfield.
From photo archives of the Marine Biological Laboratory, Woods Hole.

planned to study the hemocyanin, said in the literature to exist in the blood of the sea hare, Aplysia, a Gastropod Mollusc. But we soon convinced ourselves that the blood of Aplysia contains no hemocyanin.

Florkin and Redfield investigated instead a recently discovered echiuroid species, an odd worm-like invertebrate that strangely turned out to have red blood cells containing haemoglobin in its coelomic fluid (Redfield and Florkin 1931). These were memorable days for the young researcher: "I shall never forget our expeditions along a wild, rocky, fogbound coast, to the sands exposed at low tide where, surrounded by attentive pelicans, we spent hours, sometimes in the glorious beauty of sunrise, following the changes of oxygen content in the burrows of the worms." Next, on the spur of opportunity, they turned their attention to the respiratory properties of a sea lion: "We obtained permission to kill a specimen of sea lion, *Eumetopias stelleri*, and I keep a sad memory of our approach to one of those inoffensive animals, shooting at him with a gun and cutting its throat to collect its blood. It may be one of the origins of my lack of tendency to experiment with Mammals, and of my taste for the less human-like Invertebrates." But it only reinforced an inclination that Florkin already possessed, as these words from his autobiography make clear: "It was Léon Fredericq's example which inspired in me a lasting interest in the necessity of developing the sadly neglected field of Invertebrate Biochemistry."

As Redfield's interest in comparative physiology and biochemistry ebbed – by 1935 he had launched a brilliant career as a biological oceanographer at Woods Hole Oceanographic Institute – Florkin pursued his work on oxygen-carrying proteins, visiting the Plymouth Laboratory in England and the Concarneau station in Normandy, where his invertebrate specimens were accessible. In the process he spotted and "collected" his future wife in Concarneau. In 1934 Florkin was appointed to the newly created chair of biochemistry in the Faculty of Medicine of Liège. From that moment he was clear about his purpose: "to endeavour to bring the reign of law into comparative biochemistry, considered [until then] as a collection of scattered data, and to contribute to transforming it into a consistent and structured whole, in relation with the concepts of phylogeny and of evolution, and based on a consideration of molecular aspects."

Florkin's phylogenetic and evolutionary outlook seems on the surface to belittle the supportive role of biochemistry in assisting comparative physiology's quest to understand how animals work. By severing itself from comparative anatomy, Florkin (1979) argued, comparative physiology focused on functions while neglecting phylogenetic trees rooted in shared ancestral traits: "Comparative physiology was therefore led to compare analogous systems (respiratory systems, circulatory systems, digestive systems, excretory systems) without any reference to homologies, left to the realm of comparative morphology." In 1944 he wrote a small but important book to show how comparative biochemistry could remedy this oversight. The gestation of this book is an interesting story in itself.

The invasion of Belgium by the Germans in May 1940 forced Florkin to join the Military Hospital Service, but as the allied troops were routed he found himself trapped in the Pas de Calais region of France. To keep his mind off the stressful situation, he drafted "in a schoolboy copy-book" an essay to take stock of the literature on comparative biochemistry and organize the information in such a way as to discern evolutionary trends at the molecular level. When the Germans released him along with other physicians, Florkin was reunited with his family back in Liège. But in occupied Belgium life had changed. The heavy strategic bombing of Liège by the Allied Forces early in 1944 forced Florkin and his family to take refuge in the countryside. There he picked up the essay drafted in 1940 and expanded it into a small book, *L'évolution biochimique* (1944).

Florkin outlined the main goal of the book: "to show that biomolecules and biochemical systems vary from species to species, from genus to genus, from order to order, and in a continuous way in phylogeny." He also pointed out that biomolecules may show adaptive characters. Homology was defined as "common chemical origin" while analogy meant "a common biological activity" (Florkin 1973). In other words, what had served the comparative anatomists well in constructing phylogenetic trees based on morphological traits of animals should be applied also to the variety of biomolecules. Florkin chose to see in the variety of animal forms a reflection of the differences in their biochemistry.

In *L'évolution biochimique*, Florkin first showed that there is a chemical design common to all animals in terms of molecular constituents and reactions. This design was conserved through evolutionary time because of constraints on what is effective for producing and maintaining life. But super-imposed on this common design are more or less subtle differences among animals in the chemical structure of molecules and in what they do for the organism. The more complex the molecule, the likelier it will diverge in structure among animal groups. Florkin showed how one can distinguish analogies from homologies when comparing biomolecules. The proteins carrying oxygen in the blood, for instance, vary substantially among animal groups in their structure but, whether it is haemoglobin or haemocyanin, the protein accomplishes the same function: bringing oxygen to the tissues. So these proteins are analogous. The haemoglobins of vertebrates share fundamental similarities of structure that betray a common ancestral origin; they are homologous.

Florkin introduced also the concept of adaptation for biomolecules. Staying with the example of oxygen carriers, he found that haemoglobins adapt their affinity for oxygen (or oxygen-binding capacity) according to the oxygenation needs of the animal. A sluggish fish such as a carp exhibits a lower affinity of its haemoglobin for oxygen than an active swimmer like a trout. In the last part of the book Florkin argued that biochemical signatures are better markers of the position of animal groups in the phylogenetic tree than morphological characteristics. He also recognized that genes are behind the adaptations peppering the evolutionary history of animal groups, although knowledge of how genes manage this was to await the discovery of the structure of DNA eight years later.

Notwithstanding the fact that Florkin's book appeared in the midst of the turmoil of World War II, and that it consequently went largely unnoticed at first, its importance for the intellectual development of comparative biochemistry far outstripped that of Baldwin's *Introduction to Comparative Biochemistry*. It is arguable that Florkin elaborated the concept of evolution at the molecular level even before the advent of molecular biology. Only years later was the book's impact felt beyond continental Europe, after an English translation appeared in 1949. Meanwhile, Florkin had established an international reputation based on the dynamism of his laboratory and his numerous personal contacts with prominent physiologists and biochemists. His wide circle of connections was reflected in his 1952 review of current contributions to comparative biochemistry. This review article also showcased the talented students that Florkin attracted to his lab after the war:

Ghislaine Duchâteau, Charles Grégoire and his wife Suzanne Bricteux-Grégoire, Charles Jeuniaux, and Ernest Schoffeniels.

Florkin's influence in the biochemical field spread further through his editorial stewardship of several book series: Comparative Biochemistry (seven volumes, 1960–64), Comprehensive Biochemistry (thirty-three volumes, 1962–79) and Chemical Zoology (ten volumes, 1967–78). The latter was co-edited with Bradley T. Scheer (1914–1996), an American invertebrate physiologist trained at Scripps Institution of Oceanography in La Jolla, California, who spent his academic career at the University of Oregon. Scheer hosted a symposium on *Recent Advances in Invertebrate Physiology* and published its proceedings in 1957. Many luminaries of the field participated in the symposium, some of whom will be discussed in subsequent chapters.

One final seminal contribution by Florkin and Schoffeniels to discuss is the book *Molecular Approaches to Ecology* (1969). The incursion of comparative biochemistry into the ecological sphere must have been in the air because, as we saw in chapter 5, C. Ladd Prosser, also in 1969, was pleading for a more holistic approach to adaptive fitness which involved the molecular level. Ecologists in the 1960s viewed their field simply as the study of interactions between an organism and its biotic (organic) and abiotic (temperature, salinity, air flow, and so forth) environment. Florkin and Schoffeniels challenged them to consider as well how adaptations fit into this interactive ecosystem: "A proper approach to the study of adaptation must start from the consideration of the relation organism-environment at the level of the community or of the organism, and proceed progressively from this organismic starting point to the underlying molecular aspects." Classical ecology, the Belgians argued, exposes the interactions in an ecosystem, but the key to understanding the underlying causes of such interactions lies in the properties of biomolecules:

It is certainly true, as [Verne] Grant (1963) has pointed out, that the presence of coyotes is a factor in the determination of the number of rabbits living in a given territory. As Grant rightly states, an increase in the population of coyotes will reduce the rabbit population. The expansion of the population of coyotes will eventually stop when the number of rabbits will have reached the limits compatible with the number of rabbits they eat. No molecular approach will reveal these causal relations. Everybody will agree to this. But whatever patience and ingenuity may be devoted to counting trouts and insects living in a stream, it is only by a molecular approach and through the knowledge of the properties of trout hemoglobin that we shall understand why trouts live in streams and not in marshes, where insects also exist.

An example of biotic interaction in which molecules are involved are the amino acids dissolved in water by the decay of zooplankton. Many organisms such as sea anemones and starfish absorb these amino acids through their skin as a nutritive supplement to amino acids in prey ingested through their digestive system. Florkin and Schoffeniels revelled in providing multiple such examples. The molecular approach to ecology, in their view, helped dispel the notion that animal courtship and social behaviour are due to the nebulous instinct; pheromones and other chemical signals act with other sensory cues to shape social cohesion in an animal's ecosystem (see chapter 10).

An important aspect of Florkin's legacy was his assistance in the foundation of a new journal, *Comparative Biochemistry and Physiology*. Up until 1960 no journal dealing with comparative physiology had dedicated a section to comparative biochemistry. Even when the *Journal of Comparative Physiology*, as the *Zeitschrift für vergleichende Physiologie* became known, produced a separate section that included biochemistry, the latter had to be shared with systems and environmental physiology, and it happened only in 1984. So the opportunity for greater visibility that the new journal afforded comparative biochemistry was not to be spurned. This visibility seemed even glaring, as "biochemistry" came ahead of "physiology" in the journal title, although it is not clear if its placement signalled a hierarchy of importance or an alphabetical order.

The impetus for such a journal emerged from discussions between two quirky and controversial individuals: Gerald A. Kerkut (1927–2004) and Robert Maxwell (1923–1991). Kerkut studied at Cambridge University, where he completed a PhD in zoology on the locomotion of starfish. His contrary personality emerged early on, according to his associate Robert Walker (2004), as "his ability to antagonise resulted in his never fully achieving the acclaim many felt he deserved." As Kerkut saw no future for him in Cambridge, he took a position as a lecturer in the Department of Zoology at the University of Southampton in 1954, and by 1960 in its new

Department of Physiology and Biochemistry. He was a neurophysiologist and the research that brought him recognition concerned the electrophysiology and neurotransmission mechanisms of giant nerve cells in garden snails (Kerkut et al. 1975).

Maxwell, on the other hand, was cut from different cloth: not a scientist, not an intellectual, but an entrepreneurial British Jew whose bag of tricks, if his biographer Tom Bower (1992) is to be believed, cannot fail to recall Mordecai Richler's novel *Duddy Kravitz*. Born in a Jewish enclave in what was then the Czechoslovak Republic, near the border with Hungary, Maxwell (then named Jan Ludvik Hoch) fled in 1939 after Hitler invaded Czechoslovakia. He eventually ended up in Great Britain and served commendably in World War II, earning a Military Cross. His fluency in several languages led to his work as a field intelligence officer and the alias "Robert Maxwell," which stuck for the rest of his life. His assignment in the Public Relations and Information Services Control in British-occupied Berlin after the war built up his skills as a business entrepreneur and established a network of personal contacts that served him well on the way to the creation of his business empire in the 1950s. The manner of his business practices – doggedly and ungentlemanly according to his critics – was frequently vilified by the press and the British elite.

The jewel in Maxwell's empire was Pergamon Press, founded in 1951. It slowly started publishing scientific books and journals in all spheres of science. According to Kerkut (1988), he and Maxwell first met in 1958, when Kerkut was only thirty and Maxwell thirty-four:

> Thirty years ago, in 1958 I met Robert Maxwell in his office in Fitzroy Square, London. I was interested in finding a publisher for a book that I was writing. Robert Maxwell agreed that Pergamon Press would publish the book and we then went on to discuss a new series of monographs in zoology and also the setting up of a new journal in comparative biochemistry and physiology. I was very impressed with his positive approach and the rapidity with which he grasped the possibilities of the proposals. He was very enthusiastic and supportive and asked me to put the ideas in writing, and in the meantime to go ahead to see how things might develop.

Maxwell showed Kerkut the necessary steps: "First it was important to get a co-editor who could help get the support of his colleagues and fellow scientists. Second, an Honorary Editorial Advisory Board had to be set up. This would include eminent, active scientists who could send papers of their own and their colleagues for publication in this new journal and get it off to a good start." Kerkut obliged by selecting Bradley Scheer, Florkin's co-editor for the Chemical Zoology series, and a slew of distinguished editorial board members among whom Per Scholander, Cornelis Wiersma, and Florkin. The first issues of *Comparative Biochemisty and Physiology* appeared in 1960. From the twelfth volume forward, Florkin replaced Scheer as co-editor.

Florkin's input in the new journal constituted but one of the many avenues he followed in order to promote comparative biochemistry. At home, Liège continued to hold sway in the field thanks to his remarkable disciple Ernest Schoffeniels (1927–1992). First attracted to zoology, Schoffeniels switched to the physiology of biological membranes in the laboratory of Hans Ussing in Copenhagen (Jeuniaux and Balthazart 1993). This new expertise led to the landmark discoveries he made with Daniel Nachmansohn at Columbia University on the molecular basis of bioelectricity, using the "electroplax" of electric eels as an experimental model system. Back in Liège in the 1960s, he joined Florkin's team and followed up on Florkin's initial biochemical studies of osmoregulation. Schoffeniels is credited with discovering how cells cope with the changes of salinity and osmotic pressure in the blood when a crab moves from a marine to a more diluted (estuarine or freshwater) environment (Schoffeniels 1960; Gilles et Schoffeniels 1969). To make up for the lower salt concentration in the blood and to prevent deadly cell swelling during these osmotic stresses, the crabs mobilize "osmolytes" – small organic substances such as amino acids – to return the cells to osmotic balance.

Schoffeniels, like Florkin before him, was a prolific scientist who published several books and hundreds of articles. Ironically, this "beehive" of a man was suddenly lost to science at the age of sixty-five, when he died of anaphylactic shock after a bee sting (Jeuniaux and Balthazart 1993). As the laboratory of biochemistry had flourished under Florkin and Schoffeniels, comparative physiology was not neglected. As Florkin inherited biochemistry from his teacher Henri Fredericq, the teaching of animal physiology fell into the lap of Zénon Bacq between 1939 and 1949. (Bacq, considered a founder of comparative pharmacology, will be discussed in chapter 10.) His successor, Marcel Dubuisson (1903–1972), was trained in Ghent and moved to Liège (Bacq 1980). Dubuisson's main contributions to comparative physiology were his discovery of the synchronization of the breathing cycle with the heartbeat in several invertebrates, and the fact that the heart of some invertebrates has a nervous pacemaker thanks to a local nerve plexus, while other invertebrates do not (Dubuisson 1931). The later part of Dubuisson's research career focused on mammalian cardio-vascular physiology; and administrative duties, including the role of rector of the University of Liège, took time away from research.

In 1976 the chair of animal physiology was created at the Faculty of Science, and Raymond Gilles (1940–2018) was the first occupant. As Gilles had trained under Schoffeniels at the Biochemistry Department of the Faculty of Medicine, his transfer to physiology in another Faculty may have signalled a break of discipline. But a perusal of his scientific production suggests that he stood at the cusp of the two fields. His seamless treatment of physiological and biochemical mechanisms of osmoregulation, the traditional research topic of the Liège School, was original and brought him international recognition. His prestige was enhanced when he organized the first International Congress of Comparative Physiology and Biochemistry in Liège in 1984. With Gilles, the golden era of the Liège school of comparative physiology and biochemistry came to an end.

7

The Canadian Way:
Fish Physiology and Biochemistry

ᐤ

Fred Fry is a distinguished fish physiologist, an area of comparative
physiology to which this country of ours has made an important and
distinctively Canadian contribution.
~ Donald M. Ross (1981)

Like Belgium, Canada made a contribution to the field of animal physiology and biochemistry quite incommensurate with its size. (By size, of
course, it is population that is meant, not the spread of the country.) The
historical development of the field is closely intertwined with priorities for
the country's economic development. Surrounded by three oceans and dot-
ted with countless lakes and rivers, Canada was quick to make fisheries a
cornerstone of its commercial strategy. The way the federal government
managed the fisheries business, including quality control and research, and
the way the business affected the academic community, especially biologists, are issues that need to be put in context before examining how fishes
and other aquatic animals became staples of Canadian academic research
with a physiological or biochemical bent.

The first relevant initiative was the government's creation of a fisheries
board. Kenneth Johnstone, in his book *The Aquatic Explorers* (1977), deftly
charted the birth, growth, and evolution of the Fisheries Research Board of
Canada. Although some initiatives, private or sponsored by provincial gov-
ernments, had from the 1860s to the 1890s attempted to take stock of inland
aquatic resources of potential economic value, little attention was paid to

the marine sphere, and federal policy with regard to fisheries was conspic-
uously subdued. In the view of several academic biologists, Canada trailed
behind other countries, including its neighbour south of the border. In a letter dated 6 May 1895 to the secretary of the Royal Society of Canada, for instance, Archibald P. Knight, professor at Queen's University in Kingston,
made this plea:

> I venture to call the attention of the Royal Society of Canada to the
> desirability of having either a lake or a seaside laboratory in Canada, to
> which our naturalists could resort for some months every summer and
> undertake research work. I have myself felt the need of such an insti-
> tution, and I know of other biologists in Ontario who have felt it also.
> Last summer, for example, there were seven Canadians working at the
> Marine Laboratory at Woods Hole, Mass., and I have no doubt that
> more would have been there if they had known of the advantages of-
> fered for study and investigation. (Quotation in Johnstone 1977)

Knight's call was heeded, but even with ministerial consent and a circuitous trail of committees it took three more years to put together a
"Board of Management of the Marine Biological Station of Canada." The
grand title sounded hollow as there was no station, and only after a "portable
laboratory" had served different locations in Atlantic Canada did the permanent Atlantic Biological Station of St Andrews in New Brunswick open
in 1908 (Hart 1958). In the same year the Pacific Biological Station was established in Nanaimo on Vancouver Island. In 1912 the Board of Management morphed into the Biological Board of Canada, on which Knight
served as chair, and by 1937 it finally became known as the Fisheries Research
Board of Canada.

The staff of the stations tracked fish stocks, developed fishing gear and
techniques, worked to improve the handling, storage, and processing of fish,
and performed many more tasks within their purview. But in parallel, early
on, Knight's dream of academic biologists spending summers in St Andrews
or Nanaimo to indulge in their own research pursuits became a reality. A
person of interest in this regard is the physiologist and biochemist Archibald
B. Macallum (1858–1934). Born in Western Ontario to a Scottish immigrant
father, Macallum taught school in his late teens, saving money to afford a

university education (Leathes 1934). He graduated in zoology from the University of Toronto and in 1888 completed a PhD at Johns Hopkins University in Baltimore under physiologist H. Newell Martin. When he returned to Toronto he found himself in the same academic structure he had left in Baltimore; namely, that pending the creation of the Medical School much of the curriculum not covered in the hospital setting, including physiology, fell under the purview of the Biology Department (McRae 1987). After earning an MD in Toronto in 1890, Macallum was appointed to the just-created chair of physiology at the Biology Department. He never practised medicine.

Macallum's research interests dovetailed largely with natural history, and his paramedical physiology found an outlet only in his teaching and in the administrative reforms he put in place to secure a reputation for physiology and medicine in Toronto that went beyond the Canadian border. Sandra McRae (1987) has illustrated the initial lack of identity and self-confidence of Canadians when, under Macallum's stewardship, professors were hired from the UK but students were directed to medical schools in the United States, where they enjoyed successful careers. Macallum's own research revolved around the blood composition of inorganic substances (sodium, potassium cance. Thanks to his outstanding achievements, he was elected to the Royal Society of London in 1906, a rare accolade in those days for someone Canadian-born. It was his research drive that brought him into contact with aquatic animals and eventually to the Atlantic Biological Station.

In the second summer after the St Andrews station opened, Macallum acted as resident director while conducting research there with his assistant student Frederick R. Miller (Johnstone 1977). Miller was curious to know how nerves control the rhythmic contractions of the lobster's intestine, and his work was rewarded with a publication the following year (Miller 1910). Macallum investigated the blood composition of dogfish (small sharks), cod, pollock, and lobster. Macallum was remembered by Arthur E. Calder, the boat handler at the St Andrews station, as "a man of great dignity at all times. He also liked a drink and one of Calder's jobs was to purchase Macallum's whisky. "He never offered me a drink either. I couldn't have accepted it, for I was a teetotaler."

Calder recalled one occasion when Macallum was at the station and, after several drinks, decided that he wanted to go to St Andrews. There was only a small boat at the wharf and there was a stiff sea running. Calder warned Macallum that it would be rough going, but Macallum insisted on going just the same. Inevitably, as they rounded Joe's Point, they took one wave aboard that drenched them. Macallum received it stoically, but when they reached the wharf at St Andrews, he drew himself up and solemnly observed to Calder: "Evidently she is just a fair-weather boat" (Johnstone 1977).

We are not told if Macallum liked his drinks before taking samples of the blood of his captured animals, but the sober tone of his scientific reports suggests not. In the paper resulting from his early work in St Andrews, published in the *Proceedings of the Royal Society* (Macallum 1910), in which he investigated blood or fluid composition from jellyfish to man, Macallum stressed how the inorganic composition of the internal milieu of jellyfish and other marine invertebrates is a mirror image of that of the surrounding sea water. This means that the internal concentration of inorganic solutes is at the mercy of changing salinity in the surrounding waters. The maintenance of a constant and stable internal milieu such as observed in vertebrates, Macallum argued, "is a powerful factor in influencing the course of evolution." How? By giving "an enormous advantage to [the organism], for it can change its habitat and adapt itself to a new environment without affecting the stable conditions under which its own tissues and organs do their best work." Macallum in fact could not imagine the success of vertebrates had not their blood composition been maintained constant in a changing external environment. But what functional innovation allowed vertebrates to accomplish this? Macallum posited a response:

> The establishment of that constant internal medium would therefore appear to have been the first step in the evolution of Vertebrates from an Invertebrate form. That, on the other hand, postulates that the kidney, developed to regulate and keep constant the internal or circulatory fluid, was essentially the first typically Vertebrate organ, and therefore of origin more ancient than that of the Vertebrate brain and spinal cord.

These views, some of which failed the test of time, were later amplified in a substantial essay entitled *The Paleochemistry of the Body Fluids and Tissues* (1926). They were greatly influential in Macallum's lifetime and even inspired Lawrence J. Henderson's views expressed in *The Fitness of the Environment*

(1913) (see chapter 4). When Macallum's essay on paleochemistry was published, Macallum had left Toronto to help organize – and become the first chair of – Canada's National Research Council. This was followed with another first, as chair of the newly created Department of Biochemistry at McGill University in 1920 at age sixty-two, by which time, according to McGill biochemist Rose Johnstone (2003), "his major research work was behind him."

The man who succeeded Macallum at McGill, James Bertram Collip (1892–1965), was celebrated as a member of the team that discovered insulin, although he and Charles H. Best were left out of the Nobel Prize, which was awarded to Frederick G. Banting and John J.R. Macleod in 1923. Collip was in fact a graduate student of Macallum's and as such he participated in summer research at the St Andrews station. After his doctorate he was hired as a lecturer at the University of Alberta in 1915 and rose through the ranks there. Although Collip developed into a paramedical endocrinologist, he touched early in his career on comparative physiological aspects, travelling from Edmonton to the Nanaimo Pacific Station to get close to his animals of interest. There he studied how a host of invertebrates and fishes maintain a higher concentration of carbon dioxide in their fluids or blood than found in sea water (Collip 1920). After the discovery of insulin he returned to Nanaimo to test the effect of insulin on the oxygen consumption of fishes (Collip 1925). He found that while insulin caused hypoglycemia in the fish, oxygen consumption was unaffected.

Another member of the insulin team visited the St Andrews station. The Scotsman John Macleod was the senior member of the team and professor of physiology at the University of Toronto specializing in carbohydrate metabolism. He went to St Andrews in 1922 to investigate fishes as possible source of insulin. As James R. Wright Jr (2002) explained:

Macleod strongly believed that fish held the solution to the shortage of insulin. His belief was based on the observation that teleost (ie, bony) fish have large, anatomically discrete islet organs. These structures, unlike the microscopic islets that are scattered throughout and comprise only 1–2% of the mammalian pancreas, represented a fairly pure source of islet tissue from which insulin could be more easily extracted and purified.

John J.R. Macleod in the 1920s.
National Library of Medicine Image B018326.

From his sampling among the ugliest fishes that St Andrews could offer – eelpout, sculpin, goosefish – Macleod made his case that "insulin can be prepared by very simple and inexpensive methods from the principal islets of certain readily available fish" (Macleod 1922). Macleod was following Krogh's principle that for every physiological problem there is an animal best suited to solve it (chapter 4), a few years before it was formulated. August Krogh intrudes further in this story in that, on account of the diagnosed diabetes of his wife Marie, he visited Macleod and others in Toronto in late November 1922 "and left with authorisation from the University of Toronto to produce insulin in Denmark. During the winter of 1922–23, Krogh and his associate, Hans Christian Hagedorn, set up the Nordisk Insulin Laboratory and began production of pork insulin. Krogh also studied fish insulin but quickly decided that it was of great scientific interest but little commercial value" (Wright 2002).

Macleod so much enjoyed St Andrews – where he played golf with the director of the station, Archibald G. Huntsman (Wright 2002) – that he returned soon after the insulin expedition. He resorted to the same ugly fishes to examine the critical role glycogen plays in fish muscle contraction (Macleod and Simpson 1926). While this scientific contribution ended Macleod's incursion into the world of fishes, within two years another physiologist made extensive use of it. His story and his unsung contribution to Canadian comparative physiology deserve to be told.

~

Boris Petrovich Babkin (1877–1950) was born in Kursk in western Tsarist Russia, the son of an army officer (Burgh Daly, Komarov, and Young 1952). First attracted to music and history in his boyhood and teenage years, he shifted to medicine and, probably influenced by his father, entered the Military Medical Academy of St Petersburg in 1898. After graduating in 1901 Babkin pursued neurosurgical studies and won a Gold Medal for his medical thesis. As Babkin explained in his biography of Ivan Petrovich Pavlov (1949), the conditions under which the experimental work was performed – what with the lackadaisical attitude of his supervisor – left Babkin so disillusioned that he leaned toward trading experimental physiology for history of medicine as his future scholarly pursuit.

By a circuitous path Babkin entered Pavlov's Institute of Experimental Medicine at the Medical Academy in 1902, after Pavlov had convinced him to abandon the history of medicine and embrace experimental physiology. His first investigation under Pavlov, on the effect of soap (as emulsifier) on pancreatic secretions, went so well that Babkin determined to become a physiologist. Between 1902 and 1912, a period that includes the year, 1904, when Pavlov was awarded the Nobel Prize, Babkin was a student and then assistant in Pavlov's laboratory. It was during these years that Babkin built up his scientific strength in the physiology of the digestive system. In 1912 he was appointed chair of animal physiology in the Agricultural Institute of Novo-Alexandria (now Pulawi, Poland), and during his tenure there he published the book *Die äussere Sekretion der Verdauungsdrüsen* (The Exocrine Secretion of Digestive Glands, 1914), which earned him an international reputation. In 1915 he was called to fill the chair of physiology at the University of Odessa, where he remained until ousted from his post and exiled by the Bolsheviks in 1922 over suspicions of disloyalty.

Babkin and his wife found refuge in England. His British colleague Ernest H. Starling (1866–1927), also a pancreas expert and heralded as a founder of endocrinology, welcomed Babkin as a Medical Research Council researcher in his laboratory at University College London. However, by 1924, having been offered no permanent position, he accepted a lowly instructorship from Washington University in St Louis, Missouri. Shortly before he was to sail, however, an unexpected opening came up. In an entertaining article Babkin (1942) recalled how his Canadian connection unfolded:

Two days before our departure from London I went to say good-bye to Professor A.V. Hill, who at that time was in charge of the Department of Physiology in University College, London, where I worked for two years after leaving Russia ... We parted in a most friendly manner. I was almost at the outside door when I heard Hill rushing down the stairs two or three steps at a time.

"I just opened a letter," he said. "They want a physiologist at Dalhousie. Would you like me to recommend you?"

"Dalhousie?"

"Yes. Dalhousie University in Halifax, Nova Scotia."

Boris Petrovich Babkin.
Image derived from a group portrait by Lafayette Ltd.

"Nova Scotia? Where is that?"

"In Canada. Not far at all from here. In four or five days you can always come back to England," said the Englishman.

Babkin sailed to New York as originally planned. During the crossing he had to answer questions on a US Immigration form:

> I came to the query: "Complexion?" My complexion? Probably at that moment I was pale, but I was not sure what complexion I have in happier circumstances. After long deliberation, and being afraid of disappointing the immigration officer, I wrote down, "Complexion – all right." Thoughts about my complexion exhausted me completely. What kind of complexion had my wife? I wrote, "Complexion – pink," to which she objected very much afterwards. But I honestly thought that it would help her gain admittance to the United States.

After arriving in New York, his travails with border officials behind him, Babkin travelled directly to Halifax, where he was treated to a tour of the Dalhousie campus and the facilities he was offered. He accepted the position. "Those who have not lived through the dreadful storm of a revolution," he wrote, "can hardly understand the great relief and the feeling of profound gratitude that were mine when my battered ship entered the quiet harbour of an 'ordinary' life. It is somewhat more than pleasant recollections that I have of Dalhousie and Halifax." In the four years he spent at Dalhousie, Babkin eased into lecturing in the English language and continued his research on the mammalian digestive system. In some unexplained way he heard of a fish species (killifish) that lacked a stomach, and in 1926 he visited the St Andrews station in the neighbouring province of New Brunswick to investigate how the fish compensates functionally for this peculiarity (Babkin and Bowie 1928). This started Babkin's love affair with the Atlantic Biological Station which continued after his appointment as research professor of physiology at McGill University in 1928, and it produced fifteen papers.

The bulk of Babkin's research using St Andrews facilities concerned the skate, a fish related to sharks (elasmobranchs) but with a flat body and spread out pectoral fins that undulate during swimming. In the initial paper, Babkin (1929) explained the motivation behind his adoption of this animal model:

From the point of view of comparative physiology the skate possesses many features of interest. Animals such as the elasmobranch fishes, which are generally considered to have remained at a lower point of evolution than mammals, present an opportunity of investigating the intermediate stages of functional development of the different organs of the higher forms. An attempt was made in the present study to investigate the pancreatic secretion in skates, since the anatomical relation of the pancreatic gland in these animals affords certain advantages for such experimental study.

Babkin found that pancreatic secretion and other digestive functions in skates shared basic features with mammals but differed in details of innervation and response to chemical or pharmacological signals. He also compared the skate's circulatory system with that of mammals (Babkin et al. 1933), noting that the wall of the aorta in the skate is elastic, not muscular as in mammals. Concerning the innervation of the gut, Babkin observed: "The neuro-muscular apparatus of the gastro-intestinal tract and of the blood vessels in elasmobranch fishes exhibits a positive motor reaction to adrenaline and acetylcholine. These same two drugs, when applied to similar structures in mammals, give opposite effects."

Babkin was unusual in that he trained three women at McGill, all of whom participated in his research at St Andrews. Margaret E. MacKay (1903–2003) was born in Nova Scotia and completed both her BA and MA at Dalhousie. She followed Babkin to McGill, where she earned her PhD in 1930, the first doctorate by a woman in McGill's Department of Physiology, based on her research on the control mechanisms of the salivary glands in dogs. At St Andrews she investigated the digestive and circulatory systems of different fishes (MacKay 1929, 1932). She did postdoctoral research at Harvard and from 1939 spent her entire academic career as professor of physiology at Queen's University in Kingston, Ontario. Mary Elinor Huntsman (1910–2006) was born in Toronto, where her father was a professor in the University of Toronto's Department of Zoology. As her father was also the director of the Biological Station of St Andrews, she grew up spending her summers there. That these summers led to Huntsman's collaboration with Babkin is not surprising; she investigated the pharmacology of the skate's heart (Huntsman 1931). Her PhD was earned under the direction of Charles Best at the University of

Toronto. Finally, Helen I. Battle (1903–1994) was born in London, Ontario, where she received both her BA and MA at the University of Western Ontario. She completed her doctorate with Archibald Huntsman at the University of Toronto, becoming the first female PhD in marine biology in Canada. With Babkin she studied the digestive system of the herring (Battle 1935). She was a distinguished ichthyologist and marine biologist throughout her career at the University of Western Ontario.

∼

The era of the paramedical physiologists open-minded enough to engage in comparative physiological problems ended with Babkin. A new age of Canadian zoologists curious to know how fishes work dawned with the arrival on the scene of the pivotal and influential figures of Frederick E.J Fry (1908–1989) and William Stewart Hoar (1913–2006). Their breadth of vision and holistic approach to functional performance in fishes were legendary. They trained students and disciples who made their own mark in the next generation, and together they initiated a distinctively Canadian school of comparative physiology. Whereas Hoar may be perceived as a classical comparative physiologist schooled in C. Ladd Prosser's approach, Fry connects more with the physiological ecology practised by George Bartholomew in the United States.

Frederick (Fred) Fry was born in England but his family moved to Ontario when he was four years old (McCauley 1990). Peter Evans and colleagues (1990) relate how Fry may have arrived at his decision to dedicate his career to fish:

After World War I, the family settled in Toronto, where his father operated a wholesale fish business during the 1920s. It was through the family business and a joint venture to market frozen fish fillets that he met Dr. A.G. Huntsman, who at that time was in charge of the federal government's fisheries research laboratory in St. Andrews, New Brunswick, and the fish processing laboratory in Halifax, Nova Scotia. Huntsman, on one of his visits to Toronto, took Fred to meet Dr. W.J.K. Harkness, who was then Director of the Ontario Fisheries Research Laboratory at the University of Toronto. These connections undoubtedly influenced Fred, but when asked how he became interested in fisheries research he would say, with a characteristic twinkle in his eye, "I guess I was stubborn. I started in the fish business and just switched to research. Besides, I had already made a salmon skeleton." So it was that he changed his plans to become a chemist and instead embarked on a career in fisheries that would span more than five decades.

He studied zoology at the University of Toronto, where he obtained his BA in 1933 and his PhD in 1936. His thesis, under the direction of ichthyologist John R. Dymond, concerned the ecology of a local fish, the cisco. Around the time he became lecturer of zoology in 1938, his interest in the physiological angle of ecology was growing. At first he asked such questions as whether the amount of carbon dioxide in the water has an impact on fish buoyancy and breathing. Then, during World War II, his research shifted to aviation medicine as he served in the Royal Canadian Air Force. Back in civilian life, he returned to the Department of Zoology at the University of Toronto, this time as a tenure-track professor, staying until his retirement in 1974. For his research Fry was associated with the Ontario Fisheries Research Laboratory, which was a university facility located on Lake Opeongo in Ontario's Algonquin Provincial Park.

Fry's overriding career drive was to chart environmental factors constraining the survival and normal activities of fishes. He focused on temperature as a determinant of the level of energy metabolism. In his classic paper "Effects of the Environment on Animal Activity" (1947), Fry made a distinction between the "purely physiological approach" of analysing "the processes that go on within the organism" and classifying "the organism in the light of the activities which it exhibits." To achieve such a comprehensive physiological description, Fry classified "the inanimate factors of the environment … according to their relation to the metabolism of the organism and hence to its activity," and treated the consequences of these relationships "with particular reference to one group of organisms, the fish." How does a fish deal with the bewildering set of environmental factors he identified – lethal, masking, directive, controlling, limiting, and accessory? This is what Fry tried to decipher. How does a fish work?

Lethal factors mean those that fatally compromise metabolism by, for example, destroying enzymes. Masking factors channel the metabolic machine to support the regulation of the internal milieu of the animal. Directive factors determine which part of the habitat the animal will choose to occupy or what physiological change it will make in anticipation of an environmental change. Controlling factors can induce changes in the speed of chemical reactions, hence affecting the metabolic rate, which in turn affects heart rate, breathing rhythm, and muscle performance. Limiting factors creep in when the supply of essential materials in the animal's environment (oxygen, salts, amino acids, and so forth) does not meet the animal's needs. And finally, accessory factors are classed as being accessory to murder; if a cold temperature, for example, is not in itself lethal, it becomes so by affecting the salt-water balance of the fish.

For the first time in ecological physiology, "the total set of functional linkages between the individual fish and its surrounding world" (Evans and Neill 1990) was laid out. Fry's analysis of the subtle and incredibly diverse ways that ecological factors constrain the metabolic scope and way of life of fishes became over time the stuff of a scientific paradigm. The "Fry paradigm" has had a major impact on contemporary studies in ecology: "[T]he essence of the Fry paradigm is that life on our planet thrives near interfaces – transition surfaces (the 'patterns of forces') where the requisite conditions are most readily available to living systems" (Kerr 1990).

With his emphasis on temperature, Fry looked at how the respiratory mechanism of fishes adapts to ambient temperature (Fry et al. 1947). He and his colleagues noted that fishes that are acclimated to higher temperatures are less sensitive to carbon dioxide, meaning they can take up oxygen better even in presence of a high carbon dioxide content in water. This acclimation, they concluded, is due to the facilitating effect of temperature on blood circulation and oxygen transport in the blood. Fry further developed his thoughts on this process with the notion of temperature compensation (Fry 1958). By 1957 Fry was highly regarded by his colleagues internationally, and this reputation was reflected in invitations to review his field of investigation.

Among the forty-seven graduate students whom Fred Fry supervised, two stand out as fish physiologists who espoused their mentor's views and whose research gave credence to the Fry paradigm. J. Sanford Hart (died ~ 1974) worked with Fry on the cardiac performance of freshwater fishes (Hart 1943) and on the effect of temperature on the goldfish's oxygen consumption and cruising speed (Fry and Hart 1948a, b). In the 1950s Hart joined the National Research Laboratory in Ottawa, where he pursued research into the early 1970s on problems related to temperature management in birds and mammals, including seals. Fry's second graduate of note, J. Roland Brett (died 1991), started working on temperature tolerance in fishes in 1944, and he later reviewed the subject (Brett 1956). Around 1952 he joined the Pacific Biological Station in Nanaimo, where he remained for his entire career. He gained international accolades for his work on Pacific salmon, and in 1964 he published a landmark study of the energetics of swimming salmon which partly simulated upstream conditions (Brett 1964). Here is how Brett (1979) himself later described it: "This was the first study on the exact metabolic energy expended by a streamlined fish (salmon) for any given swimming speed within its capacity. By use of a hydrodynamically stable water tunnel, it reproduced the equivalent of the classical treadmill used for determining the energy cost of locomotion in terrestrial animals."

The project was not initially motivated by curiosity alone. Hydroelectric dams were being planned along the major salmon run of the Canadian west coast, the Fraser River. This looming urgency, Brett noted, "posed a lethal threat to the free passage of fish that annually migrate by the thousands both upriver as adults and downriver as fingerlings." With heavy investment from Canadian government research programs, Brett was able to build "a recirculating, highly controlled, tunnel respirometer," since known as the Brett-type respirometer. Its purpose was "to simulate and consequently predict the metabolic costs associated with velocity and temperature changes in river and reservoir … The tunnel included an electrified downstream grid that served to prevent any lazy behaviour from obscuring physiological capacity." His study revealed how much oxygen consumption was needed above resting consumption and at which temperature (15°C) to meet the cost of swimming upstream. The study also showed how the salmon handled fatigue.

One key contribution on this topic was Fry's chapter on fish respiratory physiology in the first book devoted to the physiology of fishes, edited by Margaret E. Brown (1957). The advent of this book is worth noting both as an important contribution to comparative physiology and as a channel for the visibility of Canadian fish physiologists.

Margaret E. Brown (1918–2009) was born in the foothills of the Indian Himalayas to a British colonial civil servant, but moved to England for her education (Pond 2009). She studied zoology at Girton College in Cambridge and after war service on a farm she obtained her PhD on the physiology of growth in the brown trout (Brown 1946). She later studied fish in Africa and Brazil and pioneered research on the husbandry of *Tilapia* as a food fish. It is not clear how she became involved in editing *The Physiology of Fishes* (1957) after she moved to Oxford University in 1955, but she points out in the preface that she intended the book to bring current knowledge on the physiology of fishes on par with that of mammals and insects, emphasizing their economic as well as scientific importance.

The book came out in two volumes. The first covered topics more or less associated with metabolism (respiratory, alimentary, excretory, cardiovascular, endocrine, and reproductive systems), and the second dealt with anything related to behaviour (nervous and sensory systems, behaviour, various effector systems, and physiological genetics). The contributors were largely British, followed by Americans (volume 2). Canadians were represented by Fred Fry (aquatic respiration), J. Roland Brett (visual system), Virginia Safford Black (1914–2001, excretion and osmoregulation), and William S. Hoar (endocrine and reproductive systems). Since Fry and Brett have been discussed above, let us here concentrate on Black and Hoar.

Virginia Black was born in New York and attended Swarthmore College in Pennsylvania. We may recall from chapter 5 that the animal physiologist Lawrence Irving held a professorship at Swarthmore. Irving took the then-Virginia Safford under his wing and in his lab she met Edgar C. Black (1908–1967), a Canadian graduate student originally from the province of Saskatchewan. (Edgar had started with Irving when the latter held a professorship at the University of Toronto, and continued studying with him when the Irving moved to Swarthmore.) From the trail left by their publication records, Edgar and Virginia married in 1941 or 1942 and the couple immediately moved to the University of Toronto, where Edgar secured a post in the Department of Zoology. Around 1946 they moved again, this time to the Department of Physiology of Dalhousie University in Halifax, where Edgar held a professorship. Finally, around 1950, Vancouver became their final home, as Edgar became the first appointed professor at the new

Department of Physiology of the University of British Columbia, and Virginia became associated with the Zoology Department.

From the period of his doctoral work with Irving, Edgar was interested in how fishes manage the transport of oxygen in the blood in the face of challenging environmental factors. In this respect he moved close to Fry's research interests. He summarized his work and reviewed the field in a paper on the respiration of fishes (E. Black 1951). Later he became interested in the physiological challenges that fishes meet when they undergo strenuous muscular exercise, as salmon do when swimming upstream. He identified dramatic rises of lactic acid and its effect on the acid-base balance in the blood as a potential cause of mortality in over-exercised fish (E. Black 1958). But he and his team found that salmon exposed to a simulated upstream migration kept sufficient glycogen stores in their muscles to handle the task (Connor et al. 1964). Edgar Black's career was cut short when he died at the age of fifty-eight. His wife, Virginia, took an interest in salmon fry that migrate from freshwater to seawater. She showed how osmoregulatory factors explained why Chum salmon fry are better adapted for fast transfer to sea water than Coho fry, which remain in freshwater for a year or more before migrating to sea water (V. Black 1951).

∿

The person who attracted the Blacks to Vancouver was William (Bill) Hoar. Bill Hoar occupied the front seat of Canadian comparative physiology not only for his research contributions but also for his seeding role in the blossoming of generations of comparative physiologists in the country. Born in Moncton, New Brunswick, Hoar pursued his undergraduate education at the University of New Brunswick (1934) and moved to the University of Western Ontario for his MA in zoology with a thesis on the development of the Atlantic salmon's swim bladder (1937). His interest in salmon was kindled by his appointment to the Fisheries Board of Canada as assistant to A.G. Huntsman over several summers, starting in 1935. Hoar was fascinated by the dramatic changes of appearance that the early stages of Atlantic salmon go through before and during their migration downstream (Hoar 1982). As he reminisced, "to me it was a major discovery and I became intrigued with the idea of finding out what was going on *inside* the fish during this [fry-parr-smolt] transformation; my curiosity was no doubt fanned by the very urgent personal problem of fixing on a suitable subject for a PhD thesis" (Hoar 1982).

Hoar was already aware of the discovery by the German-born American physiologist Frederick Gudernatsch (1881–1962) that feeding extracts of thyroid gland to tadpoles turned them prematurely into frogs (Gudernatsch 1912). From that awareness, as Hoar recalled: "My attack on the physiology of the smolt transformation was predictable. Since, at that time, I was earning my bread and butter by teaching Histology and Embryology to medical students, Bouin's fluid, a microtome, hematoxylin, and eosin were handy and nothing more was necessary to show that the thyroids of my Atlantic salmon smolts appeared histologically to be more active than those of the premigrant parr" (Hoar 1982). To achieve this, Hoar enrolled at Boston University Medical School for his PhD under the supervision of Alice S. Woodman, then professor of histology and embryology. She too became interested in salmon, publishng a study of the salmon's pituitary gland the same year that Hoar completed his thesis (Woodman 1939; Hoar 1939). The natural next step Hoar could have taken was to directly demonstrate the effect of thyroid extract on salmon parr, but he was beaten to it by a British zoologist (Landgrebe 1941).

After his doctorate and through the war years, Hoar returned to his native province and took an academic position at his alma mater, the University of New Brunswick. He took a year off in 1942–43 at the University of Toronto where his encounter with Charles Best, the co-discoverer of insulin, encouraged him to embrace the nascent field of comparative endocrinology. When Hoar was appointed professor of zoology and fisheries at the University of British Columbia in 1945, his interest in fish hormones continued, leading to the first review of this field (Hoar 1951a). In this review Hoar covered how fish hormones are involved in skin colour changes, metabolic regulation, growth, reproductive activity and behaviour, and migration. The latter two became the focus of his research activity in Vancouver. Indeed, that same year Hoar published a comparative study of the behaviour of Pacific salmon as they go through their seaward migration (Hoar 1951b). He found that Chum and Pink salmons behaved differently than Coho salmon in the way they form schools, when they are active through the day-night cycle, or move into fast water.

William S. Hoar.
Courtesy of Anthony Farrell and Colin Brauner at the
University of British Columbia.

Hoar recognized in a 1953 review not only that the theme of fish migration was important for Canadian fisheries but also that the topic is complex, as many physiological systems are engaged: growth, metamorphosis, energy metabolism, reproduction, osmoregulation, and so on. The topic also attracted attention in other countries, particularly in France, where Maurice Fontaine (1904–2009) was simultaneously conducting a research program as ambitious as Hoar's own. Fontaine, a physiologist working at what was by then the Muséum national d'histoire naturelle in Paris, first examined eel migrations (Fontaine 1944), but by 1951 he had added the Atlantic salmon to his research repertoire. In reviewing the physiology of fish migrations (Fontaine 1954), the Frenchman made the intriguing argument that eels and salmons, being among the most ancient bony fishes, had early in their evolution the same osmotic solute concentration in their blood as the sea of that geological time, so that, as the sea's salinity rose, they struggled to maintain their blood osmotic concentration at the original level against the rising ambient salinity over evolutionary time.

Hoar's years of excellent teaching at the University of British Columbia led to his textbook *General and Comparative Physiology* (1966). This textbook differed in approach from those such as Prosser's and Schmidt-Nielsen's south of the border (see chapter 5), in that it played heavily on the theme of evolution. In the preface Hoar argued that the large mass of observations of comparative physiology "are all part of the story of evolution." He wrote the text "with the conviction that a story of phylogeny in animal functions can now be sketched and that this will provide a framework into which the many details of physiology can be interestingly fitted." In this regard the title of the book's first section, "The Origins of Animals and Their Environment," set the tone. The book went through three editions and was heavily used in university classrooms in Canada and elsewhere.

By the time his textbook appeared, Hoar had trained many students who went on to make their mark in animal physiology throughout the country. He also established a strong presence for comparative physiology at the University of British Columbia's Department of Zoology, which he chaired from 1964 to 1971. He accomplished this by attracting a number of distinguished animal physiologists. One of them, David J. Randall, was a specialist in the fish cardiorespiratory system with whom Hoar co-edited a highly regarded four-volume treatise of *Fish Physiology* (1969–70). Randall, with his postdoctoral student Warren Burggren and others, went on to produce an influential textbook, *Eckert Animal Physiology* (1997), which was translated into five languages. Randall trained an impressively large number of graduate and postdoctoral students, many of whom filled professorships in several countries. Among the remarkable animal physiologists whose research impact matched that of their mentor were: Chris M. Wood, who made his career at McMaster University in Hamilton, Ontario, and dealt with the physiological impacts of environmental toxicity on fishes; UK-born Anthony P. Farrell, who passed through Mount Allison and Simon Fraser universities before developing at the University of British Columbia a fine research program on the fish cardiovascular system; and Montreal-born Steve F. Perry, whose lab at the University of Ottawa emphasized the effect of environmental stress on fish functions.

Another staff member at the University of British Columbia was Montreal-born John E. Phillips (1934–2012), who enjoyed an international reputation as an insect physiologist. Yet others are David R. Jones, who investigated how the cardiovascular system of birds and reptiles adjusts to diving and submergence, and N.R. Liley, who tried to understand how fish communicate with each other through the release of pheromones and how this mode of chemical signalling affects the ecology and behaviour of fishes. And, boding well for the field of comparative biochemistry, Hoar welcomed into his department Peter W. Hochachka (1937–2002).

∾

It is safe to assume that Bill Hoar, even if he saw great potential in the young scientist before him, could not have anticipated the brilliant international career that Peter Hochachka enjoyed. The young Albertan's burning curiosity about the ways animals adapt their metabolism to extreme living conditions, and the answers he found, created "a picture of metabolic adaptation that stands as one of the great bodies of work in comparative physiology" (Somero and Suarez 2005). Born in the province of Alberta to a family of Ukrainian heritage, Hochachka graduated from the University of Alberta in 1959 and obtained his MSc at Dalhousie University in Halifax. His very first scientific articles, as an undergraduate (Miller, Sinclair, and Hochachka 1959) and MSc candidate (Hochachka 1962) revolved around carbohydrate metabolism in trout. The latter article linked with Hoar's research interests in that it studied the effect of thyroid hormones on trout metabolism.

During his doctoral studies at Duke University, Hochachka was among the first to see in isozymes – the different forms of an enzyme coexisting in an organism – instruments of metabolic adaptation in cold-blooded animals such as fishes (Hochachka 1965). After a postdoctoral research position at the Ramsey-Wright Laboratories of the University of Toronto, Hochachka moved permanently to Vancouver, where he quickly climbed the academic ladder. Between 1966 and his untimely death in 2002, Hochachka built a research program that transformed his field, producing over four hundred articles and books, and turning out at least sixty PhD students and postdoctoral associates (Somero and Suarez 2005). What did he achieve and how did he do it?

Just a few years after his appointment at the University of British Columbia, Hochachka (1971) had arrived at a level of reflection on "the ill-defined intellectual zone somewhere between classical physiology and classical biochemistry" such that he saw fertile grounds for studying biochemical adaptation in a physiological/environmental context. In his view, nature had evolved "different forms of the enzyme [pyruvate kinase], each of which performs the same catalytic job, but each of which is integrated into, and subserves, a different physiology" (Hochachka 1971). He emphasized the distinction between properties of enzymes and their adaptational potential. "In evolutionary terms," Hochachka wrote, "two strategies are open to organisms: either to make use of properties inherent in the system under consideration, or to design, through selection, systems with new properties."

Starting from this conceptual stance, Hochachka developed a dynamic research program investigating how the metabolic machinery of cells is attuned to the wide spectrum of environmental conditions that animals such as "trout, tunas, oysters, squid, turtles, locusts, hummingbirds, seals, and humans" (Somero and Suarez 2005) must face. To implement such an ambitious program, he resorted to studies as much in the field – the Galapagos, Amazon, Arctic, Antarctic, Himalayas, Andes – as in the laboratory. It is quite aptly said that "[t]he world was both his laboratory and his lecture hall – species, lifestyles and habitats were his variables" (Suarez and Jones 2002).

Hochachka's students and collaborators derived great inspiration and stimulation from his ability to conceptualize what was at stake in his field. Colleagues who were profoundly influenced by Hochachka's guidance and mentoring, remarked that his seminal papers "frequently struck just the right balance between strongly supported arguments and adventurous flights of imagination. They were typically the sorts of 'bold conjectures' that Karl Popper urged scientists to make: concepts that are at once challenging of the conventional wisdom of a field, yet readily subject to rigorous experimental tests that might refute them" (Somero and Suarez 2005). This also meant that Hochachka had little patience with colleagues who indulged in safe "stamp collecting" – amassing data without a guiding new hypothesis.

A major collaborator of Hochachka, one who co-authored with him the great landmark of comparative biochemistry (*Strategies of Biochemical Adap-*

Peter W. Hochachka.
Courtesy of Anthony Farrell and Colin Brauner at the
University of British Columbia.

tation, 1973; *Biochemical Adaptation*, 1984), is George N. Somero. Somero's enjoyment of the cold climate of his youth in his Minnesota birthplace near the Canadian border transposed into his choice of cold adaptation as the mainstay of his physiological portfolio (Knight 2015). He left the midwest to pursue a doctorate in the laboratory of Donald Wohlschlag, a specialist of Antarctic fishes at Stanford University. He soon found himself transplanted to McMurdo Sound for ten months at a time, investigating how the metabolisms of Antarctic fishes function in waters averaging –1.9°C, and finding that the range of their temperature tolerance is very narrow, the fish dying of heat at 4°C (Somero and De Vries 1967; Somero et al. 1968)!

In 1967 Somero joined Hochachka's newly minted laboratory in Vancouver for what turned out to be a productive and enriching four years of postdoctoral studies. He acted as Hochachka's right-hand man, helping to train students and making sure the lab equipment was up and running (Knight 2015). During this period he published several papers, one of which showed that an important metabolic enzyme, pyruvate kinase, has two variants in the Alaskan king-crab, one functioning at lower water temperature and the other kicking in at higher water temperature (Somero 1969), a clear case of biochemical adaptation. Somero left Hochachka's lab in 1971 to take an academic position at Scripps Institution of Oceanography in La Jolla, California. Strangely, both Somero and his mentor were independently thinking of writing a book that sought to take stock of biochemical adaptation as a research field (Knight 2015). Repeated visits to La Jolla by Hochachka cemented their joint publication of such a book.

Strategies of Biochemical Adaptation, published in 1973, was updated to *Biochemical Adaptation* in 1984. In this major opus, Hochachka and Somero took a lofty, panoramic view of their field, emphasizing adaptation as an overarching scientific paradigm:

> When scientists attempt to take a broad view of their field of inquiry and discern the dominant conceptual themes running through their discipline, they frequently speak of the "paradigms" of the field. Such paradigms are the world-views or conceptual frameworks within which most, if not all, of the detailed questions of investigation are phrased (Kuhn 1970). In the chapters that follow we treat varied facets of what is probably the most encompassing and general paradigm in biology, a conceptual framework that finds expression at all levels of biological organization, ranging from the molecular level to the population level. This is the concept of "adaptation," the modification of the characteristics of organisms that facilitates an enhanced ability to survive and reproduce in a particular environment.

Hochachka and Somero made an important distinction between "exteriorized" biochemical adaptations such as cryptic coloration and bioluminescence, and "interiorized" biochemical adaptations (which constitute their book's topic), dealing with "the biochemical attributes of organisms that are responsible for such critical capacities as the generation of adequate amounts and types of metabolic function, the transport of gases between the cells and the environment, the maintenance of a proper solute micro-environment (pH and osmotic conditions) for macromolecular function, and the abilities to exploit the particular types of energy resources available to the organism." They took up the insightful argument of Charlotte Mangum and David Towle (1977), distinguished American animal physiologists, who pointed out that Walter Cannon's concept of homeostasis, meaning the maintenance of a physiological state, does not apply in many cases of physiological adaptation to unstable environments. Mangum and Towle coined the word *enantiostasis* to designate adaptations geared to maintain function, not physiological state.

Once these and other theoretical considerations were tackled, Hochachka and Somero asked the why question: "Why study biochemical adaptations?" They advanced the idea that "only through [the] comparative approach can the fundamental design principles of organisms be adequately understood." Using the example of a given enzyme in a rabbit, they asked which characteristics define the "rabbitness" of the enzyme. Only if variants of the enzyme were compared could the enzyme form specifically adapted to a warm-blooded mammal such as rabbit be revealed. Armed with such thinking, Hochachka, Somero, and their students were able to highlight fascinating adaptations of enzymes and other proteins to pressure and temperature in deep-sea fishes, in animals living at high altitudes or in cold climates, and in animals living without oxygen, to mention but a few examples.

This wide-ranging curiosity of Hochachka and his collaborators about functional adaptations led them to leave the comfort of their laboratories and make frequent expeditions in the field. They organized and participated in many research expeditions on the Scripps Institution of Oceanography–based research vessel *Alpha Helix* to distant ecosystems such as the Amazon and the Arctic. Other expeditions took Hochachka to the Antarctic, the Andes, and the Himalayas. This spirit was also embodied in Kjell Johansen (1932–1987), a Dane who labelled himself a Viking and a physiologist (studying the respiratory function of blood in diverse animals), who sometimes collaborated with Canadian comparative physiologists. The drive to gain "physiological insights from Nature's experiments" by roving the world is encapsulated in the title of Johansen's August Krogh Lecture, "The World as a Laboratory" (Johansen 1987). (This desire to probe the secrets of animal physiology around the world, especially seeking out animals surviving in extreme environments, will be examined in more detail in the following chapter.)

If Hochachka and Somero differed in personality – the former an extrovert, the latter somewhat shy – they both thought that doing science should be fun and conveyed this sense of elation to their students (Suarez 2002; Knight 2015). Hochachka's idea of attending scientific meetings, besides delivering ground-breaking talks, was to take the lead in partying. The profound thinker combined with the social animal to produce an irrepressible and compelling persona. Beyond his own science, Hochachka ventured to consolidate the dissemination of papers in his field. In 1994 he took over the editorship of the journal *Comparative Biochemistry and Physiology* from Gerald Kerkut (see chapter 6) with the assistance of Thomas P. Mommsen, a German-born biochemist who conducted postdoctoral studies in Hochachka's lab and later made his academic career at the University of Victoria on Vancouver Island. Under their leadership, the journal flourished and held its ground against competitors in the field. To the end, Hochachka's energy and dedication were boundless; he died from an incurable lymphoma at the age of sixty-five.

Hochachka's legacy cannot be measured only in terms of the huge impact of his scholarship and his devotion to the success of his field at large; it is evident also in the large number of outstanding people he trained. His first student, Thomas W. Moon, went on to a distinguished career at the University of Ottawa, where he and Steve Perry, one of David Randall's students at the University of British Columbia, planted the seeds for the large group of comparative physiologists who are flourishing there today. Another is William R. Driedzic, who spent his academic career first at Mount Allison University in New Brunswick and then at Memorial University in Newfoundland. Yet another was Philippines-born Raul K. Suarez, whose career revolved around the University of British Columbia and the University of California at Santa Barbara.

Perhaps Hochachka's student who best matched his exuberant and fun-loving side is Kenneth B. Storey, who made Carleton University in Ottawa his academic home. His doctoral thesis at the University of British Columbia showed how metabolic enzymes in heart muscle help red-eared turtles cope with extreme anoxia during prolonged submergence in water (Storey 1974). He went on to investigate how animals can survive freezing, and one of his pet animal models was the wood frog, which pushes the envelope by stopping breathing and heartbeat, and by moving blood to distended vessels near the heart where it solidifies during prolonged freezing (Storey and Storey 1988). This spectacular and extreme adaptation earned the amphibian the nickname "frog-sicles" in the media.

Storey's choice of experimental models (frogs, reptiles, and insects) is one of the exceptions that confirm the rule: Canadian animal physiology was and is nearly synonymous with fish physiology. There is no escaping this historical trend. But if there is a peculiarly Canadian way of exploring animal functions, its unfolding was influenced by external sources. Some of these sources originated south of the US border, although the major input came from the UK. The UK link owes much to the postcolonial Commonwealth arrangement that created natural channels for scientific exchanges. As related early in this chapter, some British physiologists relocated to Canada and established local venues for animal physiology. Later, Canadian zoologists/physiologists trained under or collaborated with many prominent British figures in the field, and collaborations extended to continental Europe.

This collaborative spirit stemmed from mutual interest in specific subfields. The question of how the gills of fishes produce the ventilation needed for gas exchange occupied the careers of George M. Hughes (1925–2011) at the University of Bristol and Graham Shelton (1930–2004) at the University of East Anglia (Randall 2014). In continental Europe Pierre Dejours (1922–2009) from Strasbourg and Kjell Johansen (1932–87) from Aarhus in Denmark dealt in the same subfield and also interacted with Canadian colleagues. The topic of how ionic exchanges take place across gills and skins and how these transports impact acid-base regulation in the blood was tackled by Jean Maetz (1922–1977) at the Station zoologique de Villefranche-sur-mer (Bornancin 1980; Kirschner 1980) and Jean-Paul Truchot at the University of Bordeaux. And the way insects deal with drought or water overload through their kidney system (tubules of Malpighi) was studied by the great pioneer of insect physiology and author of *Principles of Insect Physiology* (1939), Cambridge physiologist Vincent B. Wigglesworth (1899–1994) (Locke

1996). Wigglesworth mentored such noted Canadian insect physiologists as Kenneth G. Davey (York University, Ontario), Michael Locke (University of Western Ontario), and John E. Phillips (University of British Columbia).

While the traffic of scholarly exchanges between Canada and Great Britain reflected the strong attraction and mentoring role of the "imperial centre," there was the odd occasion of reverse traffic. A case in point is the career of Robert G. Boutilier (1953–2003). Born in New Brunswick, he completed his undergraduate studies and master's at Acadia University in Wolfville, Nova Scotia, and his doctoral studies under Graham Shelton at the University of East Anglia. His unusual talent and genial personality led to twelve primary scientific articles during his time at Acadia University alone. Boutilier became professor and department head at Dalhousie University, where his scholarly brilliance, love of fun, and interest in metabolic depression under prolonged cold exposure in various animal forms matched those of his fellow Canadian Ken Storey. Endowed with international prestige in his field, Boutilier was called in 1992 to Cambridge, UK, to take the reins of the *Journal of Experimental Biology* and pursue his academic career there (Burrows 2004). Unfortunately, a fatal illness took him prematurely, only eleven years after his Cambridge appointments.

Boutilier and his fellow Canadian physiologists and biochemists represented an uncommon breed of scientists who happened on a momentous time for a country where a broad range of factors – the lay of the land, the economic importance of fisheries, strong personalities, and the shaping of institutions – converged to drive their insatiable curiosity about the zoological landscape.

8

A Showcase of Animals
Living on the Edge
ᐁ

Our perceptions change when we see for ourselves the remotest parts
of our world, and we must change our fundamental notions of the creation
of the seafloor, of the oceans, of life.
~ Kathleen Crane (2003)

The previous chapter, through the example of Peter Hochachka and his followers, awakened us to the spirit of adventure and atmosphere of excitement
that show us how animals can work in seemingly trying circumstances. Animal physiologists willing to venture out in the world are more likely to stumble on functional adaptations that border on the exotic and titillate the
imagination. If anything, prowling the planet to its most unseemly nooks
and crannies can reinforce the experience not only of biodiversity, but more
specifically of "physiodiversity." In this chapter we follow the adventures of
animal physiologists and other biologists who dared visit these uncanny
habitats where animal functions seem to have evolved outside the box.

But should animals exposed to extreme environmental conditions be
construed as performers of extraordinary feats? Are they "super animals" on
account of special attributes such as exquisitely acute sensory functions,
overwhelming muscular strength, or sharp intellect? When addressing the
lengths to which animals go when adapting to the greatest challenges Planet
Earth can pose to their survival, animal physiologists have not succumbed
to this popular fantasy. Lynn J. Rothschild and Rocco L. Mancinelli (2001),
micro- and astrobiologists at the NASA Ames Research Center in California,

put the concept of life in extreme environments in much-needed perspective. Their starting material is the bacterial communities that live in habitats previously considered unfit for any living organism. These bacteria – dubbed "extremophiles" – seem to ride the crest of every harsh or toxic environment they find themselves in; their biomolecular make-up seems to offer all the answers to these onslaughts. Rothschild and Mancinelli could not fail to be encouraged in their belief that, if organisms on earth can survive in mind-boggling conditions, life may well exist deep in the cosmos.

They call attention to our anthropocentric tendency to associate organisms living in environments far out of our comfort zone as "lovers" of extreme environments, that is, extremophiles. Could it be that for such organisms their environment is normal and ours is extreme, even toxic? And are these organisms attracted to these so-called extreme environments or, more likely, do they simply tolerate them by evolving functional adaptations suited to the conditions? More descriptively, what have animal physiologists uncovered?

CAVE-DWELLING ANIMALS

For a start, one may ask how animals work in the subterranean world – underground waters and caves – a world where darkness and dampness prevail. Since time immemorial, caves have attracted humans as habitat, either as shelters or homes, or as art galleries of a sort, where exhibits of animal paintings can be visited to this day. But the painted or carved animals bear no affinity to the actual animals that dwell more or less permanently in cave settings. Rare cave-dwelling animals were first reported as far back as the 1400s by Chinese scholars who observed translucent fishes (Romero 2009). To Jacques Besson (1540–1573), a prodigiously inventive French engineer, we owe the first comprehensive physical survey of underground waterways in a monograph with the unwittingly humorous title *The Art and Science of Finding Waters and Springs Hidden Underground, Other Than by the Vulgar Ways of Farmers and Architects* (1569). In it Besson recorded the presence of small eels in such waters.

In the centuries following these initial observations, only larger animals such as fish and amphibians were recorded, the other cave inhabitants – mainly invertebrates – being too tiny to be noticed by amateur naturalists who visited caves. One of the largest animals found in European caves is the olm, *Proteus*, a salamander that can reach thirty centimeters in length. The olm lacks eyes and skin pigmentation. Because of its size it was the first cave animal to catch the attention of biologists. The famous French zoologist Jean-Baptiste de Lamarck was among the first to be mesmerized by the striking appearance of the salamander; and the olm, or proteus, served as an object lesson for his theory of evolution by the use and disuse of body parts and functions – Lamarckism. In his groundbreaking book, *Philosophie zoologique* (1809), he wrote:

> The proteus, an aquatic reptile akin to salamanders and inhabiting deep and dark caves in underground waters, is left with only vestiges of the visual organ … Here one comes to a decisive consideration regarding the question which I now raise. Light does not penetrate everywhere; therefore, animals which usually live in places where it does not reach lack opportunities to exercise the visual organ, if so provided by nature initially. But animals that are part of a body plan in which eyes are integral must have possessed them originally. However, since one finds some animals deprived of the usage of this organ, and which retain but hidden and covered vestiges, it becomes clear that the very shrinkage and disappearance of the said organ are the result, for this organ, of constant disuse.

Even Darwin, whose theory of evolution clashed with Lamarck's own, felt obliged to agree with Lamarck on this point and shrank from invoking the struggle for existence to account for eye degeneration in this case. According to cave biologists (biospeologists) David Culver and Tanja Pipan (2009), many cave researchers from the late 1800s until recently took a Lamarckian stance, even though the adaptationist viewpoint finally prevailed. Some pushed the envelope to the point of promoting even less orthodox theories, such as the orthogenetic theory according to which, Culver and Pipan quipped, "animals are not blind because they are in caves; they are in caves because they are blind."

The prying eye of zoologists only embraced cave research in the middle of the nineteenth century, according to the great French speleologist Albert Vandel (1894–1980), and soon the fauna of several European and North

American caves was catalogued. In his monumental book *Biospéologie: La Biologie des Animaux Cavernicoles* (1964, English translation 1965), Vandel condensed the path to experimentation in these few sentences:

> The study of cavernicolous animals has, to begin with, been carried out by amateurs with, for the most part, inadequate resources. When professional zoologists entered the field they have been, by their training, more interested in systematics and morphology than in biology and experimental research. It is only in the last few years that adequate laboratory facilities – in particular caves fitted out as laboratories – have been placed at the disposal of biospeologists.

Vandel himself was no physiologist; nor were other biospeologists until the 1950s. The German-born American ichthyologist Carl H. Eigenmann (1863–1927), in his monograph on cave vertebrates of North America, waxed eloquent on the blindness and colourless skin of cave animals, on how animals colonized cave habitats over time, and how their eyes regressed in the course of evolution, but he provided no experimental study of light perception or other sensory function (Eigenmann 1909). Such experiments, conducted by the Belgian zoologist Georges Thinès (1923–2016), showed that a totally blind cave fish from Congo – with degenerate eyes and with no optic nerve to connect to the brain – was still able to react to light by shying away from it (Thinès 1953, 1955). Thinès concluded that some light receptor other than the eyes accounted for this response, and it was later discovered that these light receptors are located deep in the brain (Tarttelin et al. 2012). The translucid skin over the fish skull allows light to reach the brain receptors.

But even if cave fish had full visual capability, eyes would be useless to navigate in total darkness. How do cave fish orient themselves without visual assistance? Among the many functional adaptations of fishes permanently inhabiting caves and listed by Thomas Poulson (1963), the heightened sensitivity of the mechanosensory system involved in distance perception and obstacle avoidance – the lateral line system – stands out (Yoshizawa et al. 2014). So do the senses of smell and taste for enhanced food detection in a habitat where food is scarce (Bibliowicz et al. 2013; Shiriagin and Korsching 2019). These compensations for vision loss are sufficient to allow cave fishes to swim safely in the dark and feed themselves. But cave fish cannot swim in the dark as fast as they could with vision, owing to the slower response of the lateral line system. This sluggishness is in keeping with their lower metabolic rate.

Other cave inhabitants of interest are bats and birds. Bats take refuge in caves during the day and fly out at night to forage for food. Like cave fishes, they too must rely on a sensory system other than visual to navigate in underground spaces. The first experimentalist to examine this problem was no other than the Italian Lazzaro Spallanzani (whose contribution to the rise of experimental biology was examined in chapter 1). The present account is based on published letters by Spallanzani (1794) and on Spallanzani's unpublished notebooks unearthed by Dutch physiologist Sven Dijkgraaf (1960). Spallanzani's interest in bats stemmed from visits he made to a cave near his birthplace, Scandiano, where he maintained a summer residence.

Spallanzani first had to ascertain that vision played no part in the bat's ability to find its way in the dark. He accomplished this in the cruellest way, by scorching the small eyes with red-hot needles; needless to say, such a practice today would not only provoke revulsion, but also fuel the outrage of antivivisectionists. He then proceeded by the elimination of other senses to finally pinpoint the ears as the source of navigation in darkness. Showing the role of hearing proved tricky, as different bat species have different auditory canal configurations that Spallanzani had to take into account to block sound reception. But how does hearing work to allow "seeing with your ears"? Spallanzani suspected that bats had to generate sounds which then bounce off objects in the cave and are reflected back to the bat's ear to gauge where obstacles lie. "As to the origin of the sound involved," Dijkgraaf (1960) noted, "Spallanzani speaks merely of 'the noise of their wings and body during flight' or (in the crawling animal) 'the sound of its walking body.' I found no indication that he ever supposed these reflected sounds to be emitted from the bat's mouth or nostrils."

The discovery that the sounds generated by bats were produced from the mouth (or nostrils in some species) emerged almost 150 years after Spallanzani's experiments. The main thread of this story originates with Donald R. Griffin (1915–2003), then an undergraduate at Harvard who was interested in understanding bat migrations. In his book *Listening in the Dark* (1958), Griffin recounted how "friends suggested that I experiment with the ability of my bats to avoid obstacles." He was put in contact with Harvard physicist George W. Pierce, who had devised a special microphone that picked up high-frequency sounds inaudible to the human ear and rendered them in a speaker as audible clicks. Together they were able to eavesdrop on bats and record these sounds, which were traced to the mouth or nostrils (Pierce and Griffin 1938). But the experimental proof that these sound emissions and their bouncing on obstacles were directly responsible for the bat's navigation in the dark was reported later (Griffin and Galambos 1941; Galambos and Griffin 1942).

Some birds also find refuge in caves where they fly freely. How do they manage in comparison to bats? Alexander von Humboldt was the first to report on the guacharo, or oil bird, in a Venezuelan cave as part of his famous *Personal Narrative of Travels to the Equinoctial Regions of the New Continent during the Years 1799–1804*. The guacharo forages for fruit at night and spends the daylight hours in the cave, where it builds nests in the upper nooks and crannies of the darkest parts of the cave. The bird can be chased by large bats sharing the cave. "It is difficult to form an idea of the horrible noise occasioned by thousands of these birds in the dark part of the cavern," wrote Humboldt (1818), "and which can only be compared to the croaking of our crows … The shrill and piercing cries of the guacharoes strike upon the vaults of the rocks, and are repeated by the echo in the depth of the cavern." Humboldt's description intrigued Donald Griffin, who wondered whether echolocation was used by the guacharo in a way similar to bats.

Griffin visited the very same cave described by Humboldt, with the difference that he brought with him the latest in sound detectors and recorders. As Griffin (1953) noted: "The most striking fact about the sounds we heard and recorded as the birds flew out from the cave was the almost complete absence of calls and screeches such as those that predominated in the roosting chambers during the day. Instead there were almost nothing but loud, sharp clicks, repeated rapidly and almost continuously as the birds flew past." Plugging the bird's ears led to collisions on obstacles, thus lending credence to the bird's use of its click sound for echolocation in the dark. The only difference between the guacharo and bats, Griffin observed, is that, unlike the sonar-like ultrasonic pitch of bats, the bird's sound is in the audible range. Similar findings obtained in cave swiftlets of Sri Lanka and the Philippines (Novick 1959).

LIFE IN THE DEEP SEA

Another habitat characterized by dim light or total darkness is the deep sea. Marine biologists make three divisions of the depth strata in oceanic zones: epipelagic (upper 200 metres), mesopelagic (200–1,000 metres), bathypelagic (below 1,000 metres), and benthic (hovering on the ocean bottom). By deep sea, we here mean the latter two zones. Like cave animals, many deep-sea inhabitants are blind or visually limited, but there is a lot more to the creatures of the deep than could be extrapolated from the life of cave animals. In fact, the diversity of lifestyles and physiological adaptations observed in the deep sea defies the imagination. Peter Herring, a distinguished British marine biologist, conveys his admiration in his book *The Biology of the Deep Ocean* (2002):

> I find the inhabitants of the deep ocean to be a constant source of surprise and delight. Every time we think we understand the ecosystem and the organisms they manage to produce a new rabbit out of the oceanic hat, so that we are required to readjust our previous perspective (picoplankton, iron limitation, hydrothermal vent communities, microscale vortex perception, red bioluminescence, phytodetritus, Archaebacteria, gelatinous zooplankton, to name a few of the rabbits).

The pioneering oceanographic expeditions of the nineteenth century awakened scientific and lay communities to the diversity and strangeness of deep-sea animals (see Anctil 2018, for a more detailed narrative). But the pioneer oceanographers only recorded their strange physiognomies; they did not address how these animals work. How could they, given the poor condition of the specimens hauled up – either dead, damaged by the trawling gear, or in the last few gasps of life – and the absence of physiological instruments best suited to answer how deep-sea animals cope with harsh environments such as darkness, cold, and hyperpressure? It was for just these reasons that marine physiologists were latecomers on the scene.

The first marine biologist who took a holistic approach to the lives of deep-sea animals, especially fishes, was Norman Bertram Marshall (1915–1996). Born near Cambridge, UK, he did not have to stray far from home for his university education and he graduated in zoology from Downing College

(Bone and Merrell 1998). After a series of excursions with fishermen aboard deep-sea trawlers, Marshall became a marine biologist. Without completing a PhD he was hired in 1937 as marine biologist on the staff of the Department of Zoology and Oceanography at the University of Hull. From the Hull homebase, "Freddy" – the nickname Marshall answered to for the rest of his life – participated in plankton surveys until 1941, when he served in war duties. In 1944 he joined an expedition to Antarctica in the capacity of marine biologist and collected marine animals there over a two-year period. In 1947 Marshall was appointed assistant keeper at the British Museum of Natural History, where he remained for twenty-five years and produced his best-known work on deep-sea fishes.

Marshall's affiliation with a museum meant that he addressed function from a comparative anatomical angle, extrapolating function from suggestive morphological peculiarities. He was no physiologist in the usual sense. But during much of his career few if any trained physiologists gave a thought to studying animals even less accessible than cave animals. Suffering from the jump of temperature and drastic decrease of pressure as they were hauled up from the deep, the irreversibly damaged animals proved unfit for experimentation. In spite of these challenges, Marshall gained numerous insights into the workings of deep-sea fishes, thanks to his holistic approach. He was interested in all aspects of life in the deep sea and for this reason he was among the pioneers of deep-sea photography to catch the denizens of the deep in their "domestic" postures and behaviour (Marshall and Bourne 1964).

An example of Marshall's approach is his investigation into the swimbladder of deep-sea fishes. The swimbladder is a fish organ designed to be filled with secreted gases "to make the density of a fish more or less equal to that of its environment so that it can swim in mid-water with a minimum of effort" (Harden Jones and Marshall 1953). Mesopelagic fishes, Marshall noted, possess well-developed swimbladders. In one of his trademark insights, he saw a link between the ability of gas-filled distended organs such as swimbladders to bounce off sound waves and the deep scattering layer (DSL) detected by boat sonars in the oceans (Marshall 1951). Originally discovered in 1942 by US Navy researchers, the DSL was considered a confusing interloper in the efforts to detect submarines or the ocean's bottom range by sonar. Marshall's observation that the recorded depths of the mesopelagic fishes coincided with the depths reported for the DSL greatly relieved the Navy; they no longer needed to treat the DSL as a direct threat to its open-sea operations.

If mesopelagic fishes have a well-developed swimbladder, further studies by Marshall and Eric J. Denton, a colleague stationed at the Plymouth Laboratory, revealed that bathypelagic and bottom-living fishes either lack a swimbladder or have one that is filled with fat instead of gases (Denton and Marshall 1958). That these fishes can still retain their buoyancy, Denton and Marshall reported, is due to remarkable adaptations. These fishes have less body mass than their mesopelagic counterparts: reduced skeleton, more fat, less protein, and more water in relation to solids. These adjustments make for more flaccid bodies and a lethargic lifestyle, as marine ecologist Eric G. Barham was later to observe (Barham 1970). In the end Marshall produced a monograph on the comparative structure and function of deep-sea fish swimbladders that has become a classic (Marshall 1960).

The sum of Marshall's thoughts on life in deep-sea habitats came out in a seminal book, *Aspects of Deep Sea Biology* (1954). This book influenced future generations of marine ecologists and also marine physiologists in particular, as technical developments opened up methods for overcoming obstacles to recording physiological activities and conducting actual experiments aboard research vessels or at seaside laboratories (Smith and Baldwin 1997). A central figure in this regard is James J. Childress. Childress received his undergraduate education at Wabash College in Indiana and pursued his doctorate at Stanford University. In 1969 he started his long-standing academic career at the University of California in Santa Barbara. It was his postgraduate research at Stanford, using Stanford's research schooner *Te Vega*, that launched him on his landmark discoveries about animal functions in the deep sea. In retrospect, if one takes into account the necessity for a man prone to seasickness to work frequently and weeks at a time aboard a research vessel, his discoveries seem the more remarkable.

Childress's first article, which appeared in the prestigious magazine *Science*, showed how a deep-sea shrimp living within a peculiar oxygen-minimum layer in the Pacific Ocean, manages to regulate its respiration (oxygen consumption) in such an oxygen-poor environment (Childress 1968). More generally, Childress observed that the respiratory rate of deep-sea shrimps and fishes is ten times lower than that of their shallow-water counterparts

(Childress 1971). Childress linked this low oxygen consumption to the very conditions of existence in the deep sea that Denton and Marshall had identified thirteen years earlier. As he explained: "The evolution of deep-sea animals apparently resulted in a combination of two characteristics ... 1) relatively large size, which deters predation, and 2) greatly reduced musculature, which decreases food needs both directly by decreasing the proportion of actively metabolizing tissue and indirectly by improving buoyancy relations." Childress went on to quantify in detail the profound changes of chemical body composition experienced by deep-sea fishes and crustaceans (Childress and Nygaard 1973, 1974).

One challenge that predatory deep-sea fishes must face in the dark habitat is the scarcity of prey and the uncertainty about when they will get their next meal – in days, weeks? Ecological physiologists wondered "how enough energy is located and acquired in such apparently depauperate environments to meet the metabolic needs of individuals," and "how energy is transferred from the productive [upper] zones of the ocean to the [lower] and deep benthic zones" (Gartner et al. 1997). Given their flaccid bodies, deep-sea fishes are poor swimmers, so they wait in ambush and use lures such as luminous "fishing poles" over their head or under the chin. They cannot be choosy, which means the prey is often larger than their stomach or themselves. It goes without saying that the gape of their mouth must be large and their stomach distensible to accommodate such prey. But there is more to their feeding mechanism than simply a wide mouth opening. The biologist who first shed light on this remarkable functional adaptation was himself a remarkable man who had to do some adapting in his own life.

Vladimir V. Tchernavin was born in 1887 in Tsarskoe Selo, where the palatial residence of the Russian imperial family was located. At the age of eighteen, and without academic credentials, he started participating in expeditions to Mongolia as "collector-zoologist" and later to Siberia, Mandchuria, and Lapland as scientific leader (Trewavas 1949). In his memoir Tchernavin (1935) recalled how it dawned on him that he needed academic training, and he entered the university, only to see his education interrupted by World War I. By 1918 Tchernavin was crippled, and with his wife Tatiana and their child endured the hardship that followed the Bolshevik Revolution. His body healed and he finally earned his doctorate "for work on structural changes in the salmon during its breeding migration" (Trewavas

1949). He was employed under the Soviet regime at the Agronomical Institute of Leningrad and managed fisheries in far regions of the USSR, especially Murmansk.

Because Tchernavin was among the fishery operators who, in the opinion of the State Police, failed to meet the requirements of Stalin's Five-Year Plan, he was arrested and imprisoned in Leningrad in 1930 for "wrecking activities" and eventually deported to the Solovetzki camp of the Gulag Archipelago. The authorities condemned him to forced labour and later allowed him to resume fishery research with no pay. In 1933, by dint of ruse and luck, Tchernavin and his family managed to escape to Finland, and the following year they moved to London, where Tchernavin started working for the British Museum. There he was handed a collection of deep-sea fishes gathered in the course of expeditions by the British oceanographic vessel *Discovery*. He became engrossed "by the functional morphology of the head of [viperfish] *Chaulodius* and the movements involved in catching and swallowing prey (Trewavas 1949). But in 1949, before he could see the results of this research in print, Tschernavin committed suicide over the untimely loss of his best friend. After surviving so many trials, the man was finally broken. His 1935 book of memoirs included the first detailed account of life in the Gulag.

Tchernavin's monograph on the feeding mechanism of the viperfish appeared posthumously in 1953. In it Tchernavin graphically described the extreme movements of the head and jaw necessary to swallow large prey. When the mouth is closed, the fang-like teeth hang outside the jaw, giving the fish a devilish appearance. As the mouth opens, two events occur: (1) the jaws and other bones at the back of the throat unfurl like pieces of a meccano toy while (2) the lower jaw is pushed forward by swinging the skull and upper jaw upward and back around the hinge with the first vertebrae. As if this was not enough to greatly enlarge the mouth opening, the gill covers are pushed outward and the heart, aorta, and gill arteries are also moved backward and downward. Thus, after the prey is impaled on the fangs, swallowing is made easier by the incredibly expanded mouth cavity. The prey, bulging in the stomach, may take weeks to digest in the frigid depths.

Tchernavin's adventure in deep-sea biomechanics inspired other studies of the same genre, and researchers soon realized that snakes living in arid habitats where food is also hard to come by have evolved similar opportunistic feeding tactics and functional adaptations for swallowing large prey.

In this regard the viperfish is fittingly named. Although the jaws and head bones of snakes are organized differently than in the viperfish, they share similar mechanical principles for engulfing extra-large prey (Gans 1961). But opportunistic living in inhospitable environments does not stop with the search for food. Finding a mate for reproduction can be as challenging, and just as they do for food, some deep-sea fishes pull out all the stops to solve that problem.

In 1925 the keeper of zoology of the British Museum of Natural History (and future boss of Tchernavin) made a most startling discovery. Charles Tate Regan (1878–1943) was educated at Cambridge University and after earning his master's was immediately hired by the British Museum, where he remained all his life (Burne and Norman 1943). Regan was credited with the first description of the exotic Siamese fighting fish. He had recently taken interest in deep-sea fishes, particularly anglerfishes, chubby-looking deep-sea inhabitants whose physiognomy ranked high in the monstrosity scale. Upon examining female anglerfish specimens, Regan noticed small fishes attached "by their nose" to the body of the female. He found out that a few years earlier an Icelander had interpreted these "appendages" as the young brood of the female (Saemundsson 1922). But on closer examination Regan realized the appendages were actually degenerated males "stitched" to the body wall of the female. Puzzled, Regan (1925) felt impelled to reach the following provisional conclusion:

> In such circumstances it would not be surprising if the difficulty experienced by the mature fish in finding mates had led the males to change their manner of life completely, in order to ensure the continuance of the race.
>
> I believe, then, that the first step was a change of habits; immature males formed the habit of attaching themselves to the females, preferably those approaching maturity, at the first opportunity that occurred. The ultimate result was that the males became dwarfed and parasitic.
>
> The structural peculiarities of the male – its small size, the outgrowths that unite it to the female, the absence of a lure and of teeth, the vestigial condition of the alimentary system – are all obviously adaptive. The evolution of these peculiarities must have been intimately related to, and even determined by, the changed activities of the male fish.

Indeed the males are almost reduced to mere sperm pouches at the disposal of females.

The American anglerfish specialist Theodore W. Pietsch has noted, however, that not all males of anglerfish families of species are obligatory parasites (Pietsch 2005). He suggests that "sexual parasitism has evolved separately at least three and perhaps five or more times within the [anglerfish group]."

Regan could not fail to notice that "[i]n relation to the darkness of their habitat [anglerfishes] are generally blackish in colour [and] their eyes are small." Deep-sea fishes are as hard-pressed as cave fishes to rely on vision. Just as the eyes of many cave fishes are technically blind, so are those of some fishes living in almost complete darkness. Interest in the visual capability of deep-sea fishes goes back to the great oceanographic expeditions of the late nineteenth century. A pioneer in this regard was the German zoologist August Brauer (1908), but it was the Danish comparative anatomist Ole Munk in the 1960s who called particular attention to the degenerate eyes of several deep-living bottom fishes (Munk 1965, 1966). Munk estimated that the young stages of these fishes have normal eyes and that the eyes regress in the course of maturation. Curiously, he found that ocular degenerative processes in deep-sea fish "show a certain resemblance to those of the retinae of mammals suffering from hereditary degeneration" (Munk 1965).

For fishes and other animals living below depths of 1,000 metres, the available residual light is generally not sufficient for them to invest in vision. But at depths of between 500 and 1,000 metres some fishes, shrimps, and squids have dug deep to adapt their eyes to visual features of their environment most relevant to their needs. Pioneers such as the aforementioned Munk (1966) in Denmark, and N. Adam Locket (1977) in the UK (later in Adelaide, Australia) uncovered the ingeniousness of deep-sea fishes in this regard. Whereas inhabitants of the greatest depths possess small or inoperative eyes, those living above 1,000 metres tend to have large eyes in order to capture as much of the available light as possible. The retina is designed for night vision, with an abundance of rod photoreceptors to enhance light sensitivity. Exceptions to this rule include rare deep-sea fishes which, seemingly against intuition, possess "pure-cone foveas" that allow them to track prey with relatively high visual discrimination (Munk 1977). And, as specialists of the photochemistry of vision have shown (Douglas and Partridge

1997), the eye is also functionally attuned to detect preferably the deep blue end of the sunlight spectrum transmitted and filtered at those depths.

When eyes are of little or no help, other sensory functions kick in. The sense of smell, for example, seems to be highly developed among deep-sea animals in an inverse relationship to eye development (Marshall 1954, 1979). Marshall (1967) was the first to call attention to the role played by the sense of smell in interactions between the sexes in deep-sea fish. He discovered, for example, that male angler-fishes have highly developed olfactory organs which are regressed in females. These allow males to detect and track female pheromone scents at a distance in the vast bathypelagic space. A sharp sense of smell, Marshall suggested, also plays a critical role in benthic (bottom-dwelling) fishes scavenging for food detritus. Marshall's hunch later led to the finding that deep-sea benthic fishes first detect by smell the general area where food is located, but use tactile sensing and taste buds on their barbels to zero in on the food's precise location (Bailey et al. 2007).

Just like cave fishes, deep-sea fishes are especially adept at detecting water displacement caused by moving animals nearby thanks to their lateral line system spread over head and body (Marshall 1979). The British ecologist Vero Copner Wynne-Edwards (1906–1997), in his influential book *Animal Dispersion in Relation to Social Behaviour* (1962), remarked that the more elongated the fishes are, the better is their ability to locate peers around them. He viewed this distribution of lateral line organs as just another tool for social cohesion:

> Attention may be drawn to the fact that few of these elongated deep-sea fish are provided with luminous organs; conversely, the luminescent species are for the most part small, short fish. This could in part be explained if these two totally different adaptations were providing alternative methods of overcoming the same problem, namely the maintenance of social contacts.

Sound production is another means of communication in a pitch dark environment. Going back to a 1857 paper by the celebrated German physiologist Johannes Müller (discussed in chapter 2), it was already known that a few shallow-water fishes generated croaking or boatwhistle sounds by "drumming muscles" stroking the swimbladder while contracting. Another of Marshall's discoveries was that in many deep-sea fishes the drumming muscles are only present in males (Marshall 1954), and it was later proposed "that males produce advertisement calls that may function in male-male interactions and attract females for mating" (Ali et al. 2016). Fishes seem to display optimal sound production and hearing sensitivity at a particular depth zone of between 750 and 1,000 metres (Priede 2017).

Ocean depths are challenging to animal life not only for their darkness and high pressure (reaching hyperbaric values as high as 1,100 amospheres), but also for their low temperatures – around 6°C in the bathypelagic zone and a uniform 4°C below 1,000 metres and on abyssal bottoms. But in 1976 a discovery by a PhD student of marine geology changed our views about thermal conditions for life on the sea bottom. Kathleen Crane, then at the Scripps Institution of Oceanography in California, and other members of an oceanographic expedition discovered a "temperature anomaly" along the Galapagos Rift in the eastern Pacific where there is a chain of submarine volcanic hills (Crane and Normark 1977). The following year Crane returned to the "hot spring" site with a team that included Robert D. Ballard and the submersible *Alvin* – the latter two famous for the discovery of the ship *Titanic* eight years later. What they found was no less than an entirely new ecosystem based on volcanic heat and sulfur in an otherwise barren and cold abyssal environment. As marine biologist J. Frederick Grassle and a host of physiological colleagues reported in 1979:

> At a time when many are looking to outer space for new forms of life, a self–contained community of unusual creatures has been discovered deep within the ocean. These animals have been found living at depths of 2,500 to 2,700 meters, in an area just over 380 kilometers from the Galapagos Islands, evoking the memory of Charles Darwin and the *Origin of Species*. The filter-feeding animals – limpets, serpulid worms, enormous clams and mussels, to name a few – are living in and around hot water vents.

To replace a food chain based on photosynthesis – unattainable in the darkness of the ocean's bottom – the animal colonies of these hydrothermal vents "are using as their ultimate food source the products of chemical synthesis – that is, the upwelling of minerals (mostly sulfur compounds) from the earth's molten interior which support a population of bacteria subsisting on hydrogen sulfide and carbon dioxide (Grassle et al. 1979). One of the physiologists who asked how these animals work was Jim Childress, mentioned earlier as a pioneer of deep-sea animal physiology. The vent organism that Childress investigated is a two-metre-tall tubeworm, *Riftia*, an odd creature devoid of stomach which obtains its nutrients from chemosynthetic bacteria cultured in a special organ, the trophosome (Cavanaugh et al. 1981).

Hydrogen sulfide in the vents must reach the symbiotic bacteria of the trophosome via the bloodstream, but sulfur compounds are usually toxic to respiratory proteins in the blood, such as haemoglobin. So how do vent tubeworms avoid sulfide poisoning? Childress and his team first reported that the sulfide binds to a specific blood protein that carries it to the trophosome (Arp and Childress, 1983), but later revised their finding when they discovered that sulfide binds to haemoglobin, albeit at a site of the molecule distinct from where oxygen binds (Arp et al. 1987). In this way sulfide intake and blood and tissue oxygenation can operate side by side without catastrophic results. As Fisher and colleagues (1989) explained, this provides for a finely tuned system whereby a sulfide-dependent carbon fixation for nutrients and energy is self-sufficient. Childress summarized his investigations and those of other laboratories in a swan song review paper (Childress and Fisher 1992).

A similar propensity for living dangerously is found also in deep-sea basins of extremely high salinity and complete absence of oxygen. Oxygen-free (anoxic) habitats may be regarded as vestiges of a long-gone past when the earth was anoxic, before photosynthesis generated oxygen in the atmosphere (Fenchel and Finlay 1995). Anoxic habitats are found in various ecosystems, and only microorganisms and protozoans have proved capable of living in them permanently. But recently a team of Italian and Danish marine biologists discovered for the first time animals that live their entire life in anoxic conditions inside sediments 3,000 metres down in the Mediterranean Sea (Danovaro et al. 2010). These are tiny animals called loriciferans, which are less than one millimetre long. Loriciferans as an animal branch were only discovered in the early 1980s, and their hiding in poorly accessible sediments may explain why they had gone unnoticed. Danovaro and his colleagues showed how these animals found solutions to the challenges of their milieu:

"(i) tolerating an enormous osmotic pressure (due to the high salinity and hydrostatic pressure); (ii) detoxifying highly toxic compounds (due to the high hydrogen sulphide concentrations); and (iii) living without oxygen."

MANAGING HEAT AND ENERGY BUDGET

If the deep sea can be unforgiving, so are deserts in their own way. Heat and aridity define them. How do animals cope in the desert ecosystem? This was a question pondered in its holistic complexity particularly by one biologist in the twentieth century. Born in British India (now Pakistan), John Leonard Cloudsley-Thompson (1921–2013) was sent to Cambridge for his college training (Griffiths 2014). When his studies were interrupted by World War II, Cloudsley-Thompson found himself serving in the North African campaign, rising to the rank of tank commander. His stint with the "Desert Rats" ended when he was seriously injured. After recuperation he participated in the invasion of Normandy. He resumed his Cambridge studies after the war and took a teaching position for some years. But the fascination that the African desert had held for him during his war service prompted him to take a professorship in zoology at the University of Khartoum in Sudan. During his eleven years in Khartoum, Cloudsley-Thompson built a reputation through his research and publications as the greatest authority on the life of desert animals.

In his crowning scholarly achievement, *Adaptations of Desert Organisms* (1991), Cloudsley-Thompson was forthright about the biggest challenge facing desert animals: "A basic problem facing all desert organisms is the maintenance of an equable temperature without using an excessive amount of water for evaporate cooling." He went on to stress the importance of behaviour designed to reach that goal: seeking shade or cooler microhabitats such as burrows, confining activity at night, and so on. But even with the use of these behaviours, there are extreme conditions in the desert that make it difficult to survive. The following is an example of a resourceful – some would call it foolhardy – animal explored by recent investigators: the Sahara silver ant.

In the 1960s a French student at the Laboratoire de Zoologie of the Université de Marseille chose the ecology, physiology, and behaviour of Sahara desert ants for his doctoral dissertation. Gérard Délye found that these insects, and especially the silver ant, are not particularly resistant to dessication and cannot survive long at temperatures reaching 40–45°C when venturing out from their nest galleries in the sand (Délye 1967). He concluded that the silver ant survives "by avoiding, thanks to its speed of locomotion, too long an exposure in the open." If the research of future investigators is any indication, Délye did not realize how understated his conclusions turned out to be.

Decades later an insect physiologist at the University of Zurich, Rüdiger Wehner, picked up the investigation. Better known for his studies of the visual performance of bees at the time, Wehner and his team turned their attention to the Sahara silver ant in the early 1980s. Amazingly, they found that "the Saharan silver ant *Cataglyphis bombycina* is exceptional in that all foragers leave their underground nest in an explosive outburst confined to a few minutes per day during the hottest midday period" (Wehner et al. 1992). The ants are on a "thermal tightrope" because, on the one hand, they cannot go out to look for food until the heat (40–47°C) forces their predator (a desert lizard) to retreat to its burrow, and on the other, they cannot stay out for more than a few minutes because the temperature continues rising up to 53°C, near their limit for survival.

When such ants forage for food, the search leads them in meandering paths over a hundred metres long but, given the time limits imposed on them by the potentially lethal heat burden they are experiencing, they cannot afford to waste time finding their way back to the nest. "How do they manage that?" asked Wehner and his team. They found that: "With a prey item in their mandibles, [the ants] return back home to their nest on a straight trajectory, instead of retracing their circuitous outbound path. The ants achieve this feat by continuously updating their home vector, which is integrated from two parameters, walking direction and walking distance of each path segment" (Wittlinger et al. 2007). The ants estimate their walking direction by a "celestial compass, reading the polarized (and spectral) sky light pattern and the sun's azimuth." The tools of the trade to accomplish this are a special part of the retina and the brain where the retinal signals are sent. Estimating walking distance and speed, on the other hand, entails computation of their strides, probably accomplished by sensory inputs from the legs to the brain (Wittlinger et al. 2007).

The desert ants can be excused for wanting more bang for the buck when they search for food on borrowed time. Siegfried Bolek and his colleagues (2012) found that ants that hit on a rich source of food preferred to return to the source location on future foraging excursions. As the ants narrowed their search with experience, they gained two advantages: they gathered more food in a shorter outing, thus limiting the risk of overheating.

Whether or not energy spent hunting for food is rewarded with sufficient returns in terms of food and energy intake is a question that has been asked since the very birth of the field of physiological ecology. Among the most fascinating examples of the "energetics" conundrum in insects are the foraging excursions of the bumblebee. The physiologist who brought this question to light is Bernd Heinrich. Born in 1940 in occupied Poland to a German landowner and entomologist, Heinrich followed his parents to the United States in 1951 and settled on a farm in Maine. In his book of memoirs, *The Snoring Bird* (2007), Heinrich chronicles the amazing and heart-wrenching trajectory of his father's family from landed gentry in what was West Prussia through two world wars and a new beginning on American soil. Of immediate relevance to us is his account of his university training and early scientific career.

Heinrich chose forestry for his field of study at the University of Maine. After his freshman year he went for a year on an expedition in Africa with his father to collect insects, and when he returned to Maine he switched to zoology. He completed his master's degree on a topic related to cell physiology, and then moved to the University of California in Los Angeles for his doctorate. He at first tried his hand at molecular biology but soon switched to a "natural history" project under the supervision of physiological ecologist George Bartholomew (discussed in chapter 5). The first important discovery Heinrich made for his PhD thesis was that the sphynx moth, a "warm-blooded" insect, maintains its thorax temperature at all times at around 42°C (six degrees warmer than humans) by transferring excessive heat through blood circulation to the cooler abdomen (Heinrich 1970). The thorax must be kept warm for the wing muscles to work effectively during flight.

Bumblebees, Heinrich noted, do not keep their thorax temperature constant at all times, but generate heat through intense thoracic muscle activity as needed (Heinrich and Kammer 1973). When a bumblebee alights on a flower to extract nectar in ambient cool air, Heinrich observed, its thorax temperature drops and the bee has to produce internal heat by muscle work before flight is again possible and it can move on to a distant flower (Heinrich 1972). This is an energy expenditure they do not need to incur if flower density is such that bees can "walk" from one to the next. Heinrich presented the cost versus reward relationship this way:

> To forage from dispersed flowers the bees must fly to each and maintain a relatively high [thorax temperature] at all times. If the reward of these flowers is small, then the bees may be unable to continue harvesting from them at low [ambient temperature] where additional energy must be expended for temperature regulation. If the droplets of nectar are minute then the bees cannot readily make an energetic profit and they may not be motivated … to expend energy either for temperature regulation or for flight.

Bumblebees have learned to develop strategies to deal with the cost effectiveness of their movements in varying circumstances. "The main idea," concluded Heinrich (2007), "was that energy economics is a key factor – the flowers must provide enough food to 'pay' the pollinator." Heinrich even wrily compared the bee world to the capitalist philosophy of Adam Smith in his *Inquiry into the Nature and Causes of the Wealth of Nations*: "In my 'capitalist bee' theory … individuals compete against each other in their foraging for nectar (and pollen). Specialization improves efficiency, which feeds into the colony's economy by providing it with more honey. No bee ever interferes with or takes anything from any other bee, except that to which all have equal access. No bee ever spreads poison or attacks any other … it showed what capitalism is supposed to be: an environment improved, not degraded, by competition" (Heinrich 2007).

KEEPING WARM

Desert ants and bumblebees represent but a small sample of animals analysed for the cost or benefit of their actions. In this respect polar regions provide as forbidding an environment for animals as deserts or the deep sea, with their own set of challenges and solutions. The Arctic fox and the Antarctic penguin will serve as examplars of problem-solving for our purpose. But first, let us step back to the notions that early investigators entertained about strategies that animals employed to keep themselves warm.

Natural historians began early in the nineteenth century thinking seriously about how birds and mammals cope with the bitter cold (Irving 1972), but it was the German anatomist and physiologist Karl Bergmann (1814–1865) who first proposed an adaptation for avoiding heat loss in these circumstances. Bergmann was born in Göttingen, where he also received his medical education and his doctorate (Anonymous 1884). He took professorships in Göttingen and Rostock, but ill health led to his premature death at the age of fifty. While teaching at the University of Göttingen, however, he published in 1847 a monograph that made him famous. In *Über die Verhältnisse der Wärmeökonomie der Thiere zu ihrer Größe* Bergmann argued the necessity "for warm-blooded animals to generate more heat per equal volume the smaller they are." Because their surface-to-volume ratio is higher than in large animals, Bergmann assumed that small animals have relatively more body surface through which to lose heat; hence the need to constantly generate more heat through shivering and other means to compensate for the loss. This became known as Bergmann's rule (Meiri 2011).

Twentieth-century investigators of course realized that strategies to maintain body temperature in Arctic conditions are more complex than Bergmann had assumed. Two important figures who set the record straight in this regard were Per Scholander and Lawrence Irving (discussed in other contexts in chapter 5). In a series of three papers they concluded that "The factors that count are the thermal properties of the surface – its insulation, its exposure, its vascularization, and its ability to tolerate a cold tissue temperature" (Scholander 1955). The Arctic fox, they found out, "needs only [a] slight increase in metabolic rate to stand the coldest temperature on earth" (Irving 1972). The reason is explained by Pål Prestrud (1991), a biologist at the Norwegian Polar Research Institute:

The arctic fox (*Alopex lagopus*) adapts to the low polar winter temperatures as a result of the excellent insulative properties of its fur. Among mammals, the arctic fox has the best insulative fur of all. The lower critical temperature is below −40°C, and consequently increased metabolic rate to maintain homeothermy is not needed under natural temperature conditions. Short muzzle, ears and legs, a short, rounded body and probably a counter-current vascular heat exchange in the legs contribute to reduce heat loss ... By seeking shelter in snow lairs or in dens below the snow cover and by curling up in a rounded position, exposing only the best-insulated parts of the body, the arctic fox reduces heat loss.

With food sources unreliable in winter, Arctic foxes move over long distances to catch prey (Lai et al. 2017) and they even cache food for survival, a practice mirrored by Polar explorers such as Roald Amundsen and Robert Scott in the Antarctic. In the Antarctic, the emperor penguin also endures hardship to secure food for adults and chicks alike. Ever since the first discovery of a breeding colony of emperor penguins by Robert Falcon Scott's Antarctic Expedition of 1901–04 (Wienecke 2010), the bird has been an object of fascination for biologists, a magnet for investigations of struggles for survival. A key investigator was the Americal polar biologist Gerald L. Kooyman. Born in Salt Lake City in Utah, Kooyman completed his undergraduate studies at UCLA and his PhD at the University of Arizona. In Arizona he started by studying the kangaroo rat (of Schmidt-Nielsen fame, see chapter 5), but somehow he was diverted to investigating the diving behaviour and physiology of the Antarctic Weddell seal, which became the topic of his doctoral dissertation in 1966. Kooyman spent his career at the Scripps Institution of Oceanography in La Jolla, undoubtably recruited by Per Scholander, the resident physiologist and expert in diving physiology (see again chapter 5).

Kooyman's interest in the emperor penguin was already manifested in 1971, when he published the first detailed study of the bird's diving behaviour when hunting prey. Here is the way he and his team explained his motives for investigatng this particular penguin (Kooyman et al. 1971):

Emperor Penguins not only dive and feed during winter darkness, as other Antarctic penguins do, but at times must do so in heavy pack ice and possibly have to do so even under fast ice, conditions that other penguins seldom, if ever, encounter. These features plus the fact that this species is the largest of diving birds make information about such diving capacities as submersion durations, depth, and swimming speed of especial interest.

Kooyman and his team found that the penguins' dives can last over eighteen minutes and reach 265 metres deep – an amazing feat. They can ascend at the astonishing speed of 120 metres per minute; and after they surface from a prolonged dive their respiration is "deep and rapid," and they begin shivering only after a delay. As they cannot take in oxygen during the dive, they are champions at storing oxygen in their muscles' myoglobin so they can continue swimming for a prolonged period deep underwater (Kooyman and Ponganis 1998).

Another investigator who made important contributions to the physiological ecology of emperor penguins is the Frenchman Yvon Le Maho. The movie *March of the Penguins* (2005) made utterly graphic the ordeal these extraordinary birds must go through to reproduce and survive: how the penguin flock walks over 100 kilometres from the water's edge to its breeding ground over a safe (thick) ice sheet; how the exhausted and starving females travel back to the water after breeding to feed while the males keep the chick warm under them in harsh winter conditions; how the females must race back to the breeding ground to feed the chick on the verge of starvation. As the ordeal unfolds some of the chicks and parents are inevitably lost. Some of the coping mechanisms developed by the penguins, as narrated in the movie, were brought to light by Yvon Le Maho and his team.

The way Antarctic penguins regulate their temperature and handle fasting was the topic of Le Maho's doctoral thesis at the Université Claude-Bernard in Lyon (Le Maho 1976). He explained how "males incubate the single egg while fasting for up to 4 mo[nths] and losing some 20 kg of their body mass"; how their metabolism is higher than average for birds; how their rigid feathers prevent heat loss in windy conditions; and how "huddling close together is essential in reducing metabolic rate" (Le Maho et al. 1976). After Le Maho established his own physiological laboratory in Strasbourg, he and his students went on to make other intriguing discoveries, such as that penguin fathers feed regurgitated food to their chick when the mother is not back in time from foraging at sea to relieve her partner (Gauthier-Clerc et al. 2000).

One question asked by the narrator of *March of the Penguins* is why emperor penguins stayed in the Antarctic as glaciation set in when many other resident species moved to milder climes. Perhaps, answers the narrator, they toughed it out because they thought the climate change was only temporary, or out of stubbornness. The same question may be asked of any organism staying put in an extreme environment. The only reasonable answer, it seems, is that if they can tinker with their physiological make-up and the conformation of their molecules to make living in harsh conditions survivable, then why not give it a try?

9

Learning How Animal Brains Work

༈

Much of my waking life and that of many of my friends is spent
racking our brains over how brains work.
~ Theodore H. Bullock (1993)

Before the intellectual environment was ripe for neuroanatomists and
neurophysiologists to develop the methodology to tackle the inner workings
of nervous systems, zoologists had already been busy investigating animal
behaviour. In the last quarter of the nineteenth century, the debate focused
on such concepts as automatic reflexes, animal instinct, and animal intelligence – concepts cast more or less in the Darwinian mold.

Thomas H. Huxley (1825–1895), the evolutionist famously labelled Dar-
win's bulldog, fired the first salvo by proposing that animals are automata
(Huxley 1874), meaning that animals are essentially machines set in motion
by predetermined sets of instructions coded in their brains. Huxley here
acknowledged his intellectual debt to the seventeenth-century philosopher
René Descartes (discussed in chapter 1). As historian of biology Robert J.
Richards explained in his book *Darwin and the Emergence of Evolutionary
Theories of Mind and Behavior* (1987): "For Descartes and his disciples, animals mimicked intelligent action, but operated as mere machines: brutes
consisted of extended matter alone and functioned according to the laws of
physics." Huxley, on the basis of his and others' experimental observations,
especially on frogs, concluded that animals are conscious (reflexive) machines, but machines nonetheless.

Someone who took exception to Huxley's assertion was no less than the man heralded as the founder of American psychology, William James (1842–1910). But as the title of James's response to Huxley, "Are We Automata?" (1879), suggests, his emphasis tilted more to humans than other animals, and his discourse took more of a philosophical turn. James was well aware that experimentalists, who rely on physical evidence, and animal psychologists, who look for evidence of a mind at work, are two solitudes that rarely meet. In objecting to Huxley's argument, James cited the idiosyncratic powers of the brain's mental correlate: "Over a frog with an entire brain, the physiologist has no such power. The signal may be given, but ideas, emotions or caprices will be aroused instead of the fatal motor reply, and whether the animal will leap, croak, sink or swim or swell up without moving, is impossible to predict." This indeterminate quality, James adds, is particularly pronounced in animals possessing cerebral hemispheres, that is, mammals and primates particularly.

William James envisaged that consciousness, seen as an emanation of brain activity, "slowly evolved in the animal series, and resembles in this all organs that have a use." He believed that out of the dual Darwinian "mechanical processes" of natural selection and spontaneous variation consciousness can emerge, although these processes "can in no sense be said to intend it." If Darwinian mechanisms do not intend to bring about consciousness, then what does? James seems to invoke an emergent property of the brain and a case of directed evolution:

As [an animal's] body morphologically was the result of lucky chance, so each of his so-called acts of intelligence would be another; and ages might elapse before out of this enormous lottery-game a brain should emerge both complex and secure. But give to consciousness the power of exerting a constant pressure in the direction of survival, and give to the organism the power of growing to the modes in which consciousness has trained it, and the number of stray shots is immensely reduced, and the time proportionally shortened for Evolution. It is, in fact, hard to see how without an effective superintending ideal the evolution of so unstable an organ as the mammalian cerebrum can have proceeded at all.

A central figure who picked up the debate where James left it was George Romanes (1848–1894). In the book *Dawn of the Neuron* (2015) I discussed in detail his life and his contribution to understanding the role of nervous systems in shaping the behaviour of simple animals such as jellyfish. But Romanes's scientific interests went beyond jellyfish, and his deep friendship with Charles Darwin, almost forty years his senior, oriented his incursions into physiological evolution at large and the evolution of animal intelligence in particular. In his book *Animal Intelligence* (1882), Romanes justified his intellectual pursuit by noting that "within the last twenty years the facts of animal intelligence have suddenly acquired a new and profound impor-tance[;] from the proved probability of their genetic continuity with those of human intelligence, it would remain true that their systematic arrange-ment is a worthy object of scientific endeavour."

Romanes's dual aim was to provide a comparative analysis of animal "psychology" and to interpret the cognitive abilities of animals in the light of Darwin's theory of descent. But how does one decide that an animal is endowed with such abilities when current tests of intellectual capacities were designed for humans with whom one can interact by agreed-upon communication channels like language? Romanes trod carefully on this minefield, pointing out vaguely that the activities of organisms are the "am-bassadors" of mental ability. He added that an animal's activities "must be of a kind to suggest the presence of two elements which we recognise as the distinctive characteristics of mind as such – consciousness and choice." He cautioned that not all apparent choices made by animals necessarily re-flect a mind in operation.

Romanes's take on William James's distinction between predictable reflex responses and indeterminate mental operations was couched in Darwinian language:

> Objectively considered, the only distinction between adaptive move-ments due to reflex action and adaptive movements due to mental per-ception, consists in the former depending on inherited mechanisms within the nervous system being so constructed as to effect particular adaptive movements in response to particular stimulations, while the

latter are independent of any such inherited adjustment of special mechanisms to the exigencies of special circumstances.

So Romanes's guiding criterion for distinguishing an animal mind at work is: "Does the organism learn to make new adjustments, or to modify old ones, in accordance with the results of its own individual experience?" He also tackled the cultural tradition that opposed an animal's instinct to human rational power:

> In popular phraseology, descended from the Middle Ages, all the mental faculties of the animal are termed instinctive, in contradistinction to those of man, which are termed rational. But unless we commit ourselves to an obvious reasoning in a circle, we must avoid assuming that all actions of animals are instinctive, and then arguing that because they are instinctive, therefore they differ from the rational actions of man. The question really lies in what is here assumed, and we can only answer it by examining in what essential respect instinct differs from reason.

Romanes indeed claimed that the essential difference between reflex and instinct is that the latter involves consciousness, while reason upstages instinct in that it involves the power to perceive analogies or deduce "inferences from a perceived equivalency of relations." Rooted as Romanes was in evolutionary and neurophysiological perspectives, for him it meant that "[the] advent and development of consciousness, although progressively converting reflex action into instinctive, and instinctive into rational, does this exclusively in the sphere of objectivity; the nervous processes engaged are throughout the same in kind, and differ only in the relative degrees of their complexity." Whereas instinctive behaviour is an adaptive response to situations previously experienced, he adds, "rational actions are performed under varied circumstances, and serve to meet novel exigencies which may never before have occurred even in the life-history of the individual." To Romanes a seemingly intelligent performance by an animal cannot be affirmed as such "until an animalcule shows itself to be teachable by individual experience."

After taking all these caveats into account and reviewing the literature critically, Romanes reached the conclusion that animals at the level of earth-

worms and lower display dubious or no instinct. Higher invertebrates such
as insects clearly possess it. Lower vertebrates, in his view, show less intelligent behaviour than insects. His verdict on mammals was that there is an
enormous range of intelligent ability depending on taxonomic groups. But
only in primates did Romanes find evidence of "reasoning power," although
his evidence was often anecdotal and short on rigorous experimentation.

In a subsequent book, *Mental Evolution in Animals* (1883), Romanes asked
to what extent the state of development of the nervous system reflects mental
ability in different animal groups. He soon found out that neuroanatomical
knowledge was so deficient in his time that he was forced to rely on a crutch:

> Now it admits of being abundantly proved that throughout the animal
> kingdom, so long as we regard the muscular system as our index of the
> structural advances taking place in the nervous system, we find this
> index to consist in the growing complexity of the muscular system, and
> the consequent increase in the number and variety of co-ordinated
> movements which this system is enabled to execute.

Making inferences about the level of organization of the nervous system
from muscular development should have proved, even to Romanes, too re-
ductive a process. The brain is as much about sensory processing and com-
puting information as about controlling muscles downstream. But to
obviate the necessity of a crutch, a determined effort had to be made to understand the neuroanatomical and functional intricacies of the brain and
the peripheral nervous system. A scientific program of that nature was well
on its way in Romanes's time. One is reminded of the pioneers of the cerebral localization of function by the likes of Paul Broca, David Ferrier, Eduard Hitzig, and Hermann Munk in the 1860s and 1870s (Finger 2000).
Another breakthrough was the microscopic visualization of interconnected
nerve cells and their insertion in layered cerebral arrays by Santiago Ramon
y Cajal, thanks to the histological method of Camillo Golgi (Finger 2000).
But these milestone contributions, even though the experimental subjects
included a variety of mammals, were directed largely at understanding the
human brain and its pathologies.

One venue that took a broader, curiosity-driven view of animal brains and
behaviour was the *Journal of Comparative Neurology*, founded in 1891. Its

founder, Clarence L. Herrick (1858–1904), was a naturalist from the American midwest who turned to neurological research to fill the void on the subject in North America at the time (Windle 1975). Eager to start a scientific journal that would cater to comparisons of structure and function of the nervous system from fish to humans, with an ultimate goal of tackling the mind-body problem, but with no help, financial or otherwise, coming to the fore, he took up the burden on his own. As William Windle explained: "With characteristic audacity, the *Journal of Comparative Neurology* was launched in 1891 as a private enterprise, without consultation or advice. (There were no colleagues in the vicinity)."

The journal at first published primarily neuroanatomical work interpreted in terms of animal habits and motor functions, but functional studies on sensory and motor skills soon crept in, and the objects of study occasionally extended to invertebrate animals. The journal was off to a good start and attracted attention in Europe, but Herrick suffered a massive pulmonary hemorrhage due to overwork and had to abandon the editorship of the journal in 1894, even though his name remained on the masthead until his death in 1904. The new managing editor was Clarence's younger brother, Charles Judson Herrick (1868–1960), who at twenty had not yet earned his PhD. Charles went on to a brilliant career in vertebrate comparative neuroanatomy at the University of Chicago, where he trained many future American neurologists (Bartelmez 1973). One important change that he made to the journal in 1904 was to attract a co-editor in the person of Robert M. Yerkes (1876–1956), a noted expert on animal behaviour who studied primates.

The early issues of the *Journal of Comparative Neurology* ran the whole gamut from the hypertrophied taste centres of the catfish brain (Herrick 1905) to the hypothesis that the striate body of the mammalian brain is the seat of consciousness (Carus 1894). Early in the twentieth century, Charles Herrick was positioned at an enviable hub of scientific activity regarding the budding field of "neurobiology," and he reflected hard and deep on the evolution of the nervous system and behaviour. In his book *Neurological Foundations of Animal Behavior* (1924), forty years after Romanes's 1883 essay, Herrick could not boast that the subject had progressed: "the whole field of mental evolution, including the problem of psychogenesis, has so far remained almost impenetrable."

⌁

The field of neurobiology was at a crossroads. Correlating brain organization with animal behaviour failed to satisfy the investigator curious to know how animal brains work or to understand the intricacies of behaviour in an animal's natural environment. The brain was understood as an intricate web of lobes, tracts, histological layers, and the like. Behaviour was superficially reduced to a series of animal "habits." To break this conundrum, two paths of investigation were traced, starting in the 1930s. One research program was designed to select animal models in which nervous structures are simple enough that the activity of individual nerve cells or small groups of nerve cells can be monitored and their impact on sensory and motor functions can be directly measured. It was a neurobiologist's way of heeding Krogh's principle (discussed in chapter 4): "For a large number of problems there will be some animal of choice or a few such animals on which it can be most conveniently studied." Another research program arose that selected animals with tractable behavioural traits – predation, courtship, and so on – that were amenable to experimentation to gain insights into social behaviour. These two programs rarely intersected, forming two solitudes that persist to this day. We will now examine these two paths in turn.

Charles Herrick clearly recognized in 1924 that studying brain cells is key to understanding neural mechanisms and their evolution in the animal world, but he was powerless to implement such a program. For a start, he and many of his generation were prisoners of a narrow comparative approach that precluded concentrating on a specific animal model, and they were hindered by their focus on vertebrate animals, whose nervous systems contain too many small nerve cells that were proving intractable for electrophysiological recording at the cellular level. They also missed out on the technical innovations that allowed one to record with microelectrodes the activity of specific nerve cells. The earliest examples of the new approach are the squid giant axon and the crayfish nerve-muscle preparations.

We noted in chapter 4 how J.Z. Young serendipitously discovered the squid giant axon at the Naples Zoological Station. Young showed that larger nerve axons conduct excitation faster, and are therefore ideal for fast escape reflexes (Young 1985). He and colleagues such as Martin J. Wells – son of G.P. Wells and grandson of the famous novelist H.G. Wells – also discovered how different lobes of the octopus brain are involved in learning and memory. His discovery made the squid an unparalleled experimental model for neurophysiology. Not only is the squid giant axon a crucial brain cell for mediating the squid's fast escape response to the presence of predators (Young 1938), but its sheer size also made it a convenient tool of fundamental neuroscience for understanding how a neuron's action potential works, a discovery that led to the award of a Nobel Prize to Alan L. Hodgkin and Andrew F. Huxley in 1963.

At this juncture it seems appropriate to dwell on a recording method that not only allowed Hodgkin and Huxley to achieve their milestone work on the squid, but also allowed future neurobiologists to successfully exploit other invertebrate nervous systems where large neurons are found: intracellular recording. Ever since Emil du Bois-Reymond's pioneer recordings of bioelectrical activity (see chapter 2), physiologists had recorded from muscles or nerves what amounted to mass acticity – compounded activity from action potentials of numerous constituent cells. This changed when the British physiologist and Nobel laureate Edgar D. Adrian (1889–1977) made the innovation of using extracellular electrodes (capillary electrometres) in conjunction with a special "valve" amplifier to record the rapid electrical activity of single nerve fibres (Adrian 1925).

But extracellular recordings could not sense what goes on inside a nerve or muscle cell, or follow the dynamics of the differential electrical charges building up across the cell membrane. To probe these, one had to penetrate the cell with an electrode tip small enough to avoid permanent damage and filled with a saline solution conducive to recording reliable and undistorted electrical signals. The glass microelectrode proved to be the needed innovation. "The historical account," Allen Bretag (2017) noted, "shows that progress in developing and using the glass micropipette electrode was haphazard with numerous inventions and reinventions, with advances and regressions, with missed opportunities and false starts, and with both mistaken and correct interpretations of results." Many prototypes emerged from the 1920s on, but none did the job satisfactorily until Hodgkin and Huxley's breakthough in 1939. The tip of their microelectrode was 100 micrometres in diameter, too large to penetrate most nerve cells which have a smaller diameter. So Hodgkin and Huxley were grateful when J.Z. Young at the Plymouth Marine

Laboratory alerted them to the giant squid axon, 500 micrometres in diameter. Hodgkin and Huxley (1939) explained why their discovery was such a breakthrough:

> In the first place they prove that the action potential arises at the [membrane] surface, and in the second, they give the absolute magnitude of the action potential as about 90 [millivolts] at 20°C. Previous measurements have always been made with external electrodes and give values which are reduced by the short-circuiting effect of the fluid outside the nerve fibre.

In the 1940s the preparation of microelectrodes improved so that smaller tips of 5 micrometres or less were consistently produced and smaller excitable cells such as frog leg muscle cells could be penetrated (Graham and Gerard 1946). This capability opened the door to widespread use of the technology starting in the 1950s, but the very small size and difficulty of accessing vertebrate nerve cells ensured that invertebrates steamed ahead.

Concurrently with the squid axon, another invertebrate model emerged in the 1930s, thanks to a figure who actually proved more critical than J.Z. Young in pioneering the new cellular neurobiology. Cornelis A.G. Wiersma (1905–1979) was born in The Netherlands and studied under Hermann J. Jordan, who, as we saw in chapter 3, had established a chair of comparative physiology in Utrecht (Florey 1990). If "Kees," as Wiersma was later nicknamed, had seen the remarkable drawings of the crayfish brain and peripheral nervous system published by the famous Swedish neuroanatomist Gustav Retzius in 1890, he would have been inspired to record the activity of the large neurons stained with methylene blue that seemed to jump right off the plates. The doctoral thesis that Wiersma produced in 1933 provided just that inspiration. Wiersma was blessed with technical skills, but he needed to broaden his palette of research tools. So in 1931–32 he spent time in England to study in Cambridge with the famous neurophysiologist Edgar D. Adrian (1889–1977), who was about to win the Nobel Prize, and to compare marine species of crustaceans at the Marine Biological Laboratory in Plymouth with the crayfish he was working on.

What did Wiersma's doctoral research reveal? Wiersma stimulated motor neurons of crayfish and other crustaceans and, with refined extracellular sil-

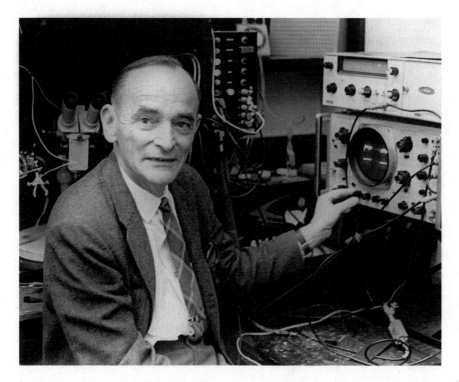

Cornelis Wiersma at his workstation.
Courtesy of Caltech Archives.

ver electrodes, recorded "muscle action currents" in response. He revealed how, contrary to the nerve-muscle systems of vertebrates, which are ruled by simple excitation, crayfish motor neurons can produce slow or fast muscle contractions depending on the way they are excited; and he showed how another set of motor neurons can only cause suppression of muscle contraction (inhibition). Oddly, it seemed that the neuromuscular system of the crayfish is more complex than that of vertebrates, while the crayfish brain is simpler than that of its its vertebrate counterpart.

How swiftly Wiersma's scientific life changed after his doctorate is an interesting story in itself. Thomas Hunt Morgan (1866–1945), the genetics pioneer whose team at Columbia University elucidated the mechanisms of inheritance in fruit flies, had been appointed in 1928 to the California Institute of Technology (Caltech) to create a Division of Biology (Sturtevant 1959).

Morgan had complained for years about what he considered the stranglehold of morphologists on American biology, and Caltech, where no biological tradition yet existed, offered the perfect venue for the reforms he had in mind. Just as he was winning the Nobel Prize in Physiology or Medicine in 1933, Morgan was scouting for comparative physiologists to follow up on his reforming zeal. He held Hermann Jordan in high esteem and asked him if he knew of prospective candidates for these open positions at Caltech. Jordan suggested, of course, his former students Wiersma and A. van Harreveld, the latter Wiersma's collaborator for years to come (Florey 1990).

At Caltech Wiersma continued research on the neuromuscular system of the crayfish but a decade later turned to the central nervous system. Neuroscientists have credited Wiersma with two major discoveries that "changed the ways we think about what nervous systems do and how they are organized" (Mulloney and Smarandache 2000). First, Wiersma (1947) showed that the firing of an action potential in any of four giant axons in the nerve cord of the crayfish is necessary and sufficient to produce the fast tail flip, which is the signature of the crayfish's escape behaviour. He later called neurons that execute this kind of task *command neurons* (Wiersma and Ikeda 1964), and the name has stuck to this day. The other breakthrough discovery is that of the brain's central pattern generator (CPG). Wiersma, working with Graham Hughes on the crayfish (1960), and the American neurobiologist Donald M. Wilson independently on the locust (1961), found that disconnecting the sensory input from the limbs to the brain does not affect the rhythmic cycle of limb movements for their locomotion – swimming in crayfish, flight in locusts (Mulloney and Smarandache 2000). This finding indicated the presence in the brain of a circuitry (the CPG) that self-generates a motor program for locomotion.

This kind of investigation, which called for an understanding of neuronal organization to account for a specific behaviour, became the field of neuroethology. Ethology – the science of behaviour – emerged as a specialized scientific discipline in the 1930s, but of course scores of natural historians have reported observations of animal behaviour since antiquity, starting with Aristotle who described the dance of the bee but failed to understand its meaning (Egerton 2016). The word "ethology" was invoked in the past to signify the science of ethics (Chavot 1994), but its use in the context of animal (and human) behaviour was pioneered by the French zoophysiologist Georges Bohn at the dawn of the twentieth century (as mentioned in chapter 3). What caused ethology to take off revolves primarily around one man, Konrad Lorenz (1903–1989).

Konrad Zacharias Lorenz, the son of a wealthy orthopaedic surgeon, was born in Vienna. Throughout his childhood, "Konrad kept both tropical and local fish species, amphibians and reptiles as well as invertebrates from the Danube" (Krebs and Sjölander 1992). He apparently developed in childhood the method of observation that served him well in his professional life. "Much of his scientific work was based on observing small groups of tame or semi-tame animals living in and around his house or laboratory" (Krebs and Sjölander 1992). Lorenz enrolled in medicine at the University of Vienna to placate his father but took a major in comparative anatomy to satisfy his zoological interests. He learned from one of his professors how to reconstruct animal genealogical trees by comparing anatomical features of sets of animals. He was later to apply this principle to behavioural traits (Krebs and Sjölander 1992).

In his book *The Foundations of Ethology* (1981), Lorenz alludes to his dissatisfaction in the 1930s with two opposing theories of behaviour heralded by "doctrinaire" psychologists – the overriding importance of purposeful instinct *versus* compulsory associative learning or behaviourism – as an impetus for him to come up with explanatory theories grounded in biology instead on the basis of detailed and careful observations. By embracing the comparative approach, Lorenz was able to discern common patterns in the execution of animal movements that in his mind reflected a common, ancestral origin of innate behaviours called fixed action patterns (Lorenz 1937). He was reassured about the validity of his theory by the experimental work of the German neurophysiologist Erich von Holst (1908–1962).

As early as 1932 in his published doctoral thesis at the University of Berlin, Holst had described ingenious experiments that he conducted in order to demonstrate that earthworms do not execute their undulating, crawling locomotion by a chain of reflexes from one body segment to the next, as the behaviourists claimed, but rather by the spontaneous generation of rhythmic activity that travels within the nerve cord from head to tail (Holst 1932; Lorenz 1962). He then turned to fishes and found a similar oscillating pattern of activity in their spinal cord to account for the coordination of their swimming and breathing movements (Holst 1934). This finding led in turn to the

Konrad Lorenz.
From the entry for Konrad Lorenz in Encyclopaedia Britannica.

startling discovery that a brain oscillator network imposes its tempo upon the spinal cord oscillator – what Holst called the magnet effect (Holst 1936). The meeting of minds between Lorenz and Holst, which occurred in 1936, led to the notion that endogenous pacemakers in the brain are at the core of behavioural physiology.

But the professional and personal relationships that developed between Lorenz and Holst did not translate into close interactions between classical ethologists, as represented by Lorenz and the Dutch Nikolaas Tinbergen, and neuroethologists at large. The questions asked by classical ethologists – what stimulating factors initiate behaviours? how are individual and social behaviours organized? and how do they change during development and evolved over time? – tended to be holistic; those of neuroethologists, on the other hand, flirted with reductionism – how an assembly of neurons works to initiate and coordinate behaviour. In short, ethology, unlike neuroethology, distanced itself from the preoccupations of physiologists. Neuroethology evolved from Holst's approach – involving surgical transections of nerve paths and electrical stimulation of the vertebrate brainstem – to the cellular approach pioneered by Wiersma. Thus the fundamental notion of the brain's central pattern generator, however inchoate its expression in Holst's texts, became rooted in the reality of specific neurons in the work of Wiersma and his followers.

One pivotal neuroethologist who followed in Wiersma's footsteps was Theodore Holmes Bullock (1915–2005). Bullock found inspiration in Wiersma's work, as evidenced in his vibrant homage to the pioneer of cellular neuroethology (Bullock 1977), but he went beyond Wiersma's cellular approach by taking a broad comparative approach to the study of animal brains in order to understand their evolution. Ted Bullock was born in Nanking, China, to American missionary parents (Zupanc and Zupanc 2008). When he was thirteen his family moved back to the United States and he received the remainder of his schooling in California. He attended the University of California at Berkeley, his father's alma mater, where he graduated in zoology in 1936. His PhD program, pursued in Berkeley, resulted in publications on the functional organization of the nervous system in acorn worms, a little known taxonomic group of marine species remotely related to vertebrates (Bullock 1940, 1944).

Through his appointments as a postdoc at the Yale School of Medecine (1940–44), assistant professor at the University of Missouri School of Medicine (1944–46), professor of zoology at UCLA (1947–67), and professor of neuroscience at UCSD and the Scripps Institution of Oceanography in La Jolla, Bullock and his collaborators went from one momentous discovery to another: recording signal transmission at the squid's giant synapse (Bullock 1946); using intracellular microelectrodes to reveal the pacemaker and co-ordinating activities of the nine large neurons composing the innervation of the lobster heart (Hagiwara and Bullock 1957); discovering a new mode of communication between neurons by electrical, not chemical, synapses (Watanabe and Bullock 1960); discovering that electric fishes also possess sensory receptors geared to detect their own electric discharge (Bullock et al. 1961); discovering that the facial pit organ of vipers is a thermal (infrared) sensor to detect nearby warm-blooded animals, to name a few.

In his autobiography Bullock (1996) examined his approach to the science that produced so many great milestones, noting his "penchant for the relatively neglected issue, technic, or animal group and avoidance of the popular one." He insisted on "the need for more comparison of taxa, particularly the phyla and classes representing major grades of complexity of brains, and the need for descriptive exploration of the phenomenology they manifest – natural history, in the best sense." For him the comparative approach was paramount. A case in point is the giant axons of invertebrates and fish, a neuronal fixture that promised to provide a more tractable physiological model to understand basic neural principles than the small neurons of mammals:

In the 1930s and for decades thereafter, the giant fibers of earthworms, crayfish, squid, and many teleosts were nothing more than an extreme specialization for some advantage, like an elephant's trunk or tusks. We focused on giant fibers as accessible cellular units, hoping their membrane and synaptic properties were not too specialized to teach us general physiology. Each had had its dramatic history of discovery and debate as to whether it was vascular, supportive, or neural. My own interest was not so much in the cellular and membrane mechanisms as in the organization of the afferent and efferent system and the integration at giant synapses.

The earthworm's marine relatives, polychaete annelids, were inter-esting for other reasons, mainly because of the extreme diversity, among families, in the development of giant fibers and of the nervous system as a whole. The diversity made them the most valuable group for arriv-ing at a plausible view of the biological meaning and behavioral cor-relates of giant systems, with confirmation from work with crustaceans, cephalopods, teleosts, and others, including odd groups like phoronids and lungfish ... It appeared that giant fiber systems are not so much es-cape mechanisms as startle response devices and that saving time by fast conduction is not as important as synchronizing a widespread mus-culature.

If the diversity of giant axon systems can lead to the discernment of a com-mon function such as a startle response in the above example, Bullock cau-tions that notwithstanding commonalities one should be alert to what the variety of brains can teach us:

A conclusion I defended in an essay in *Trends in Neuroscience* [1986] is that differences found between taxa are as important as commonalities, in understanding how brains work and how life should be understood. Nature has provided two great gifts: life and then diversity of living things, jellyfish and humans, worms and crocodiles. I don't undervalue the investigation of commonalities but can't avoid the conclusion that diversity has been relatively neglected, especially as concerns the brain.

One potential reward of revealing brain diversity through comparative analyses is the evolutionary insight to be gained. Bullock attached enormous importance to this aspect, as supported by his many publications touching on the evolution of the nervous system. In his autobiography he rejoiced in the promise held by such studies:

This is the fortunate fact that the profoundly complex brain of humans is biologically continuous with brains of simpler organisms. Further-more, many animal groups in the long evolution through invertebrates and vertebrates have been so successful that living species reflecting some of the stages are available. If we apply adequate methods, they can reveal the vast span of grades of intricacy between non nervous protozoans and sponges, nervous but brainless jellyfish, flatworms with a simple brain, insects, octopus, lizards, parrots and the host of other groups that manifest the principal achievement and least studied side of evolution – the great development of brains and behavior.

Although Bullock paid his respects to mainstream neuroscience and in return earned the respect of mammalian neuroscientists, his zoology-oriented strategy for understanding how brains work could not help but make him feel like the proverbial "underdog." In fact, he encouraged the younger generation of neuroethologists to follow in his "underdog" traces, to embrace neglected topics, anticipating that such topics would emerge as the motors of new significant knowledge in neuroscience (Bullock 1996). The extent to which his call was heeded bewildered him (Bullock 1977): "I recall vividly, but with a strange feeling in retrospect, the catalogs of 160 identified sensory neurons and afferent interneurons (INS) (Wiersma 1958), and of another 75 (Wiersma and Hughes 1961) in the crayfish central nervous system. While the elegance of these findings was well appreciated at the time, little did most of us realize that they presaged the great boom in identified cells, still nearly 10 years away."

The gathering wave of interest in – and pleas for – what invertebrate nervous systems had to offer was superbly articulated by Stanford-based neurobiologist Donald Kennedy (1931–2020) in a 1967 *Scientific American* article. Kennedy, who was to become in turn head of the Food and Drug Administration, president of Stanford University, and editor-in-chief of the magazine *Science*, articulated the problem facing neuroscience in those years:

The nervous system of a man comprises between 10 billion and 100 billion cells, and the "lower" mammals men study in an effort to understand their own brains may have two or three billion nerve cells ... These vast populations of cells present a formidable challenge to biologists trying to understand how the nervous system works. Since the system is made up of cells, one would like to understand it in terms of cellular activities, and by examining the activity of single nerve cells investigators have been able to learn a great deal about the nature of the nerve impulse and about the generation and transmission of the patterns of impulses that constitute nervous signals.

Kennedy went on to explain how neuroscientists studying mammalian brains, far from being inclined to think outside the box, resorted instead to ingenious ways of circumventing the challenges they faced by relying on biochemical studies and computer-generated models of brain waves. "The trouble," he diagnosed, "is that most of these methods treat cells as anonymous members of a population rather than as interacting individuals." Enter the invertebrate brains, which "are so economically built that for certain functions one may hope to specify the activity of every individual cell." But can one infer that tapping this knowledge provides the necessary leverage to understand how mammalian brains work? Yes, answers Kennedy, if the complexity of the mammalian brain is not due to its neurons acquiring more sophisticated capabilities than invertebrate ones but rather, as the evidence suggests, because the greater number of neurons allows for a greater diversity of interconnections between neurons. "Therefore," he concludes, "an understanding of the connections underlying behavior in a simple system can lead to useful conclusions about the organization of much more complicated ones."

~

The boom of identified neuron–based neuroethology to which Kennedy and Bullock allude took place in the early 1970s, at which time the field became known as neurobiology, "a discipline devoted mainly to cellular mechanisms underlying behavior" (Clarac and Pearlstein 2007). But as the nerve cells of selected invertebrates were the only ones large enough for reliable recordings of their electrical activity with the techniques available fifty years ago, and the simple behaviour of these invertebrates was relatively easy to analyse and interpret, neurobiologists essentially flocked around invertebrate preparations. This new brood of neurobiologists recoiled, however, from looking at their work as mere curiosity-driven research. The hope, as expressed by neurophysiologist Stephen W. Kuffler (1913–1980), a Hungarian-American thought by many to be the pillar of modern neuroscience, and his British colleague John G. Nicholls (1929–) in their book *From Neuron to Brain* (1976), was that "examples from a lobster or a leech will have relevance for our own nervous system."

At the time of Kuffler and Nicholls's writing, a lot was being learned from the nervous system of leeches and lobsters (Clarac and Pearlstein 2007). Perhaps the simplest nervous system used as a model for the study of the control of behaviour is that of the leech. John Nicholls actually initiated the use of the leech model in the 1960s with a few collaborators. In a review article Nicholls and his student David Van Essen (1974), explained disarmingly what sets the leech model apart – a description that fits other invertebrate neurobiological models:

> The leech has only a small number of nerve cells performing any particular function. For example, one sensory cell innervates a large patch of skin that in the mammalian nervous system would be supplied by many cells. As a result one can selectively activate a single sensory nerve cell by natural stimulation and thereby set up a simple reflex. The chain of events that this stimulus to the animal sets in motion can then be followed sequentially through the nervous system, again by looking at individual elements.

Nicholls and Van Essen were able to "geolocate" neuronal cell bodies in the ganglia of each body segment and give each one a name. In this way they proceeded to map much of the whole nervous system. To find out what cell or cells each recorded neuron makes contact with, and thereby figure out their function, all the neurobiologist had to do after profiling the neuron's electrical response was to inject a dye into it with a microelectrode and follow the trajectory of the dyed axon and its branches all the way to their cellular targets. Before long they elucidated the neural control at the cellular level of several leech activities: heartbeat, local body bending, body shortening, swimming, crawling, and feeding (Kristan, Calabrese, and Friesen 2005).

Nicholls also used leech neurons in culture to find how contacts between neurons lead to the formation of synapses for signal transmission. Major developments in this field, also using the leech model, were achieved by one of Nicholls's post-doctorate trainees, Pierre Drapeau (Haydon and Drapeau 1995), at McGill University's affiliated Montreal General Hospital. Leeches, used for medical treatment (bloodletting) since ancient Egypt, had made a comeback in hospitals for neuroscience research. But these worms have since remained in hospitals for other purposes: as anticlot agents after plastic, reconstructive surgery, for instance (Renault 2019).

Fascinating discoveries have also been made by poking the lobster's nervous system. We have seen how Bullock and his team worked out the simple circuitry of the lobster's cardiac ganglion, but an even more important lobster model emerged: the stomatogastric ganglion, which controls the activity of gastric muscles. It all started with Donald M. Maynard (1929–1973), a Harvard-educated zoology student who completed a PhD thesis at UCLA on the very same lobster cardiac ganglion just mentioned (Maynard 1953, 1955). Maynard happened on the stomatogastric ganglion ten years later when posted at the University of Michigan, recording the activity of some of the thirty-five neurons making up the ganglion (Maynard 1967, 1972). He showed how the interactions between these neurons could account for the rhythmic properties of the ganglion's output. In short order two investigators, Allen I. Selverston in San Diego, California, and Maurice Moulins (1936–1995) at the zoological station of Arcachon near Bordeaux, picked up the promising lead from Maynard, who died prematurely in 1973.

Selverston and his collaborators (1976) explained why studying the lobster's stomatogastric ganglion can be immensely rewarding to a neurobiologist:

> One form of behavior that is particularly amenable to physiological analysis is repeated, rhythmic movement. This is a ubiquitous form of behavior, found in almost all the phyla which have been examined. Rhythmic movement is necessary for behaviors such as respiration, walking, swimming, flying, mating, and chewing. A system that is ideally suited to the study of rhythm generation is the stomatogastric ganglion of the spiny lobster (*Panulirus*).

The activity of each neuron in the ganglion's circuit can be monitored and the various manifestations of the stomach's behaviour easily quantified. Selverston even added a new tool of his own making: selectively knocking off one neuron and watch what it does to the circuit: "Single neurons can be easily and rapidly killed by filling them with the fluorescent dye Lucifer

Yellow and illuminating them with intense light. Using this technique, the complexity of neural networks can be reduced, allowing more quantitative investigations into the mechanisms of neuronal integration than are possible with intact systems" (Selverston and Miller 1982). In this way the complete wiring diagram of the circuit involved in chewing by the gastric mill and dilatation/contraction of the pylorus was mapped.

Maurice Moulins had initiated his investigations of the lobster stomatogastric system in the 1970s (Moulins et al. 1974), but his original contributions emerged in the early 1980s. This happened thanks to Robert M. Robertson, a postdoctoral fellow in his laboratory who went on to conduct original research on the insect nervous system at the University of Alberta, McGill University, and Queen's University in Kingston, Ontario. Robertson and Moulins (1981) found that neurons from an upstream ganglion, the commissural ganglion, acted as command oscillators on the stomatogastric ganglion, entraining the latter's rhythmic frequency to that of the commissural command neurons.

Another Canadian who made an indelible mark in crustacean and insect neuroscience is Harold L. Atwood, born in Montreal in 1937. His father being an entomologist, it is not surprising, in the words of his famous sister, Margaret Atwood (2018), that "collecting bottles, test tubes, and insects on pins were no strangers to Harold." He studied zoology at the University of Toronto and, in keeping with his ability as a child to capture crayfish (M. Atwood 2018), he completed his PhD on crustacean muscles at the University of Glasgow in 1963. He conducted postdoctoral studies at the University of Oregon and at Caltech under Cornelis Wiersma, mentioned earlier. A mentor of great importance to Harold was Graham Hoyle, both his PhD supervisor in Glasgow, where Hoyle was lecturer, and his postdoctoral overseer in Oregon, where Hoyle was professor from 1960 to his death in 1986. Before examining Harold Atwood's career and achievements, it is appropriate that we look into Hoyle's own distinguished (and controversial) career.

Graham Hoyle (1923–1986) was at the centre of research mapping identified insect neurons involved in the motor behaviour of insects. He was introduced to intracellular microelectrode recording by José del Castillo (1920–2002), a Spanish neurophysiologist then working in Bernard Katz's lab at University College London. (Katz was later to win the Nobel Prize in Physiology or Medicine with the Swedish Ulf von Euler and American Julius

Axelrod in 1970 for their role in elucidating how neurotransmitter substances are released from nerve endings and handled by the postsynaptic cells.) Hoyle's collaboration with del Castillo resulted in a paper in which they showed that neuromuscular transmission in the locust jumping leg – as reflected by the production of muscle cell postsynaptic potentials upon motor nerve stimulation – is no different than its counterpart in the frog leg neuromuscular system (Del Castillo, Hoyle, and Machne 1953).

Hoyle went on to apply the same methodological approach involving intracellular recordings to crustacean neuromuscular transmission in Wiersma's lab at Caltech (Hoyle and Wiersma 1958). After Hoyle moved to the University of Oregon, he and his students embarked on numerous and original studies showcasing the structural and functional diversity of invertebrate muscles, and linking identified interneurons and motoneurons with behaviour, especially in insects and crustaceans. Among these contributions one can mention: the neural basis for the flight mechanism of locusts (Hoyle and Burrows 1973); the modes of neuromuscular activity in insects (Usherwood 1977); and even the mapping of neurons involved in molluscan swimming (Willows, Dorsett, and Hoyle 1973).

Hoyle received more than his share of attention not just for his remarkable research but also for his unorthodox views, which he expressed in acrimonious tones reflecting what many saw as his belligerent personality. At the root of his resentment was his perception of a condescending attitude on the part of medical and cellular physiologists toward zoophysiology, an issue frequently highlighted in this book. Zoology, as he saw it, was underestimated by physiologists and biophysicists: "Physiological zoology was something you did as a first project, or if you were not quite up to doing cellular biophysics. It never occurred to the majority of physiologists that comparative physiology should serve routinely as the testing ground for all supposed 'general' theories" (Hoyle 1976). And the expression of his derogatory attitude toward the usefulness of vertebrate models for neurobiology seemed articulated to in turn incense vertebrate neuroethologists:

A majority of neurobiologists are going to continue to work on vertebrates because, by reason of training or inclination, they have little or no interest in invertebrates. It is unlikely that they will be able to give

comprehensive answers to the questions of neuroethology of vertebrates, but nothing will divert them from doing what they can with
backboned animals. The vertebrate knowledge is going to be in a com-
partment of its own, and I see very little value, given the present state
of our understanding of integration, in seriously attempting to compare
functioning of invertebrate with vertebrate nervous systems, or to pretend that we can use the former as models for the latter. (Hoyle 1975)

Not content to be dismissive about the value of vertebrate neuroethology,
Hoyle pushed the envelope by claiming that only the pursuit of arthropod
(especially insect) models can bring a comprehensive understanding of the
neural mechanisms subtending behaviour. Two years before his death Hoyle
encapsulated his thoughts on the future of neuroethology, a move that in-
curred the wrath of a large segment of the neuroethological community,
who responded by condemning his dogmatism:

The initial aim of neuroethology should be to examine the neurophysiological events in a variety of behaviors, exhibited by diverse animals
from different phyla, which meet the criteria of innate behavioral acts.
The behaviors should be sufficiently complex to interest ethologists, yet
they should be addressable with neurophysiological methods down to
the cellular level. In the case of vertebrates this may mean working with
brain slices as well as whole animals, but for some invertebrates recording should be possible in the nearly intact animal during execution of
the behavior. The work will be exacting and very difficult, and it is not
likely to get done at all unless neuroethologists recognize that they
should both train and discipline themselves and restrict their attention
to well defined goals. (Hoyle 1984)

Harold Atwood begged to differ with his old mentor on at least one point.
Whereas Hoyle found it futile to pretend that invertebrate nervous systems
could be useful models for explaining vertebrate nervous systems, Atwood,
according to his former student Milton P. Charlton (2018), "pointed out how
crustacean neuromuscular junctions were useful models for events in the
vertebrate [central nervous system]. This was a frequent theme in the intro-

ductions and discussions of his papers." These contributions came out in his capacity as professor at the Department of Zoology (1965–81) and chair of the Department of Physiology (1981–91) at the University of Toronto.

In Atwood's research lab his team "used a combination of ultrastructural and physiological analyses to provide an understanding of how nerve cells and their synapses respond to experience and activity" (Dason, Sokolowski, and Wu 2018). Using two arthropod models – crayfish and fruit fly – Atwood zeroed in on the complexity and diversity of neuromuscular synapses. His team discovered how a muscle's response to repeated stimulation can improve in amplitude and readiness for more than one hour – what they called synaptic facilitation (Sherman and Atwood 1971). They observed "silent" synapses, which can be recruited into action in the course of this synaptic facilitation (Wojtowicz, Smith, and Atwood 1991). They also found that synapses can adapt their activity when there is a sustained change in the pattern of stimulation to which they are subjected – an example of long-term adaptation or plasticity (Llenicka and Atwood 1985).

One pivotal invertebrate preparation, whose implications for long-term adaptation and plasticity illuminated our understanding of learning and memory, was the sea slug (or sea hare) *Aplysia*. The story begins as it did for other invertebrate neurobiological models: the sea slug has large and identifiable neurons highly amenable to intracellular recording of their electrical activity. The setting was France, a country where neurophysiology lagged behind the UK in the first decades of the twentieth century (Barbara 2007). Angélique Arvanitaki (1901–1983) was born in Cairo from Greek (and francophile) parents and moved to Lyon to pursue her university education (Ternaux and Clarac 2012). A brilliant student, she decided to undertake doctoral studies under physiologist Henri Cardot, a specialist in rhythmic activities in heart and breathing functions. Cardot was then also director of the Tamaris marine station in the south of France, founded by his predecessor in Lyon, Raphael Dubois (see Anctil 2018, chapter 8). In Tamaris, Arvanitaki elected electrophysiology as her approach and in 1938 produced a thesis on nerve responses in a variety of invertebrates available in the bay facing the marine station (Ternaux and Clarac 2012).

After earning her doctorate Arvanitaki took an interest in the sea slugs in Tamaris. In contrast to the giant axon of the squid, what attracted her in the sea slug was the strikingly large nerve cell bodies in its ganglia and she quickly realized that studying them would show how a central nervous system works in a simple, accessible preparation (Arvanitaki and Cardot 1941). She was able to observe that many of these ganglionic neurons produce spontaneous, rhythmic spike potentials and that the rhythmic activity gradually becomes synchronized between the adjacent neurons. Unfortunately, the death of her mentor Cardot in 1942 and the occupation of Tamaris by the Italians put an end to Arvanitaki's research on *Aplysia* (Ternaux and Clarac 2012). She resumed in the 1950s, but by then another researcher had entered the fray and taken over the leadership in *Aplysia* research from Arvanitaki. His name was Ladislav Tauc (1926–1999).

Born in what was then Czechoslovakia, Tauc studied plant physiology and wrote a thesis on plant bioelectricity (Israel 2000). In 1948 he moved to France and joined the team of the famous neurophysiologist Alfred Fessard at the Institut Marey in Paris. In 1954 Fessard alerted Tauc to Arvanitaki's work and it quickly prompted Tauc to adopt the sea slug as his experimental model (Ternaux and Clarac 2012). Endowed with great technical skills, Tauc was able to insert two microelectrodes into *Aplysia* giant neurons, injecting current with one and recording graded or action potentials with the other. In this way he charted the dynamics of the neuron's activity at different locations in the cell (Tauc 1957). His most important claim to fame, however, is that he introduced Eric R. Kandel to the *Aplysia* model.

Kandel was born Erich Kandel in 1929 in Vienna and left for the United States in 1939, as his family suffered from Nazi persecution (Kandel 2006). After studying history and literature at Harvard he switched to medicine at New York University and became acquainted with brain electrophysiology in Harry Grundfest's lab at Columbia University. There he came to realize that to understand psychological processes he "would have to learn how to listen to neurons, to interpret the electrical signals that underlie all mental life" (Kandel 2006). At the same time (1956–57) he was made painfully aware of the limitations that mammalian neuroscientists experienced in asking minute brain cells to reveal their secrets, and of the apparent ease with which other neurobiologists could find answers by poking large invertebrate neurons. "I had no specific idea in mind," he recalled, "but I was beginning to think like a biologist. I appreciated that all animals have some form of mental life that reflects the architecture of their nervous system, and I knew I wanted to study nervous system function at the cellular level. All I knew at this point is that someday I might want to test an idea with an invertebrate animal."

Meanwhile Kandel embarked on investigations of the hippocampus, considered a seat of learning and memory in the mammalian brain ever since discoveries by McGill University neuropsychologist Brenda Milner (Scoville and Milner 1957). Kandel had always been fascinated with the memory process, but his electrophysiological work on the hippocampus proved intractable at the cellular level. To Kandel it represented a dead-end and it led him to search for an animal that, in MIT neurogeneticist William (Chip) Quinn's quip, would have no more than three genes, be able to play the piano, and learn these tasks with a nervous system containing only eight large neurons (personal communication)! Kandel came across an animal presenting a somewhat less winning version of the sought-after animal, when he attended lectures by Arvanitaki and Tauc in United States around 1960. He resolved to spend postdoctoral time in Tauc's Paris lab to acquaint himself with *Aplysia*.

Tauc, who had been skeptical at first that the secrets of learning and memory functions could be revealed in simple experimental models such as *Aplysia*, taught Kandel the electrophysiology relevant to the sea slug and let his visitor use the facilities of the marine station of Arcachon, near Bordeaux, where sea slugs are readily available. In 1962–63 Kandel produced results suggesting that converging synapse activity at a giant neuron is enhanced by pairing a test stimulus to one nerve pathway with a priming stimulus to another nerve pathway – what he called heterosynaptic facilitation (Kandel and Tauc 1964). If these results hinted at one elementary form of learning in an isolated ganglion, Kandel realized that he needed to find a simple reflex function in intact sea slugs for which the neuronal circuitry can be worked out and by which the full features of learning and memory can be revealed. Once he accepted a professorship at the School of Medicine at New York University, his alma mater, he implemented this plan.

As of 1967 Kandel gradually assembled a formidable team of collaborators who over decades helped him achieve his goal – unravelling the brain mechanisms of learning and memorization at the cellular, circuit, and molecular

Eric R. Kandel.
Photograph by Bengt Oberger.

levels. His efforts earned him the Nobel Prize in Physiology or Medicine in 2000. His team first mapped all the identifiable neurons in the abdominal ganglion of *Aplysia* that control the behaviour of interest to them (Frazier et al. 1967). Next they selected *Aplysia*'s gill-withdrawal reflex – a protective response to perceived threats – as their model behaviour to study learning. As Kandel (2006) explained:

In 1969 I was joined by Vincent Castellucci, a delightful and highly cultivated Canadian scientist with an extensive background in biology who regularly trounced me in tennis, and by Jack Byrne, a technically gifted graduate student with training in electrical engineering who brought the rigor of that discipline to bear on our joint work. Together, the three of us identified the sensory neurons of the gill-withdrawal reflex. We then discovered that in addition to their direct connections, the sensory neurons formed indirect synaptic connections with motor neurons through interneurons, a type of intermediary neuron. Those two sets of connections – the direct and indirect – relay information about touch to the motor neurons, which actually produce the withdrawal reflex by means of their connections with gill tissue. Moreover, the same neurons were involved in the gill-withdrawal reflex in every [sea slug] we studied, and the same cells always formed the same connections with one another. Thus, the neural architecture of at least one behavior of Aplysia was amazingly precise.

Thus they worked out the neuronal circuitry involved in facilitation as a learning mechanism (Castellucci et al. 1970, 1976). Kandel's team went on to decrypt molecular mechanisms for long-term memory and associative learning (Kandel 2006). Their accomplishments hinged on having Aplysia shipped to New York and surviving in tanks filled with artificial seawater. Traditionally researchers had worked with marine animals at seaside laboratories, but the 1960s saw a shift to inland maintenance and husbandry of marine experimental animals, thanks to technological developments. Kandel obtained his sea hares from Rimmon C. Fay (1929–2008), who supplied Southern California marine specimens to countless researchers. (Included among them was this author, who stood at the receiving end of Fay's services for many years.) After receiving an order, Fay would set out to dive and collect the requested specimens, place them in plastic bags filled with oxygenated seawater and transfer the bags into Styrofoam boxes with ice. The boxes were whisked to Los Angeles International Airport and shipped by direct flight to destination. With few exceptions the specimens arrived in excellent condition.

Informed by his experience with Aplysia, Kandel went on to explore the mammalian brain where now he knew where to look with more sophisticated tools than were available earlier. His interest in the new field of "molecular cognition" led him to new insights into the role of the hippocampus. Thanks to him and myriad other neuroscientists, our understanding of the mammalian brain at the synaptic circuit and molecular levels has snowballed ever since. This approach turned out very well for explaining a specific behaviour in terms of experimentally tractable brain circuits, but what of higher cognitive functions involved in "mind" activities?

∼

Early in this chapter we followed attempts by George Romanes to define intelligence as opposed to instinct and to assess its level in the animal kingdom. As neuroscience progressed in the century after Romanes's death and it became clear that the questions neuroscientists could answer by rigorous experimentation narrowed down considerably, the question of animal intelligence was deemed off-limits for neurophysiologists and became the purview of psychologists and philosophers. One notable exception of a neuroethologist who has meditated at length on the topic of animal minds was Donald R. Griffin, the very same whose pioneer research on bat echolocation we discussed in the previous chapter. Griffin embraced the question of "animal awareness," as he called his exploration, late in his career, as he explained in his autobiography (Griffin 1998):

> Maturing scientists often experience what might be called the "philosopause" as they turn to more general questions than those that have occupied their attention for many years of detailed investigation. Mine has involved a growing dissatisfaction with the reductionistic viewpoints so prevalent in biology and psychology. In particular, I had begun to doubt the wisdom of totally ignoring the possibility that animals may experience conscious thoughts and subjective feelings. This led me to attempt to launch a subdiscipline of cognitive ethology.

Griffin's first book dealing with the sensitive subject was *The Question of Animal Awareness* (1976, revised 1981). It has the trappings of both essay and pamphlet, pleading for the existence of animal consciousness based on little experimental, but plenty of circumstantial, evidence. Basically, Griffin built his case by emphasizing that some animals display flexible adjustments of behaviour when exposed to new challenges and sophisticated modes of communication among their mates or with other species. He also stressed that a large body of neurophysiological evidence points to common basic operating principles in all central nervous systems. He developed his thesis further in *Animal Minds* (1992) and other publications. Over the years his controversial views aroused a barrage of criticism, with the less negative ones allowing that a simple form of cognition exists in animals – as best demonstrated by Kandel and other research teams – but not consciousness. "Reopening questions about the private, subjective experiences of animals," Griffin (1998) lamented, "has aroused considerable opposition from some psychologists and ethologists. One of my books (Griffin 1984) has been called 'The Satanic Verses of Animal Cognition.'"

The field of cognitive ethology acquired a renewed theoretical impetus with the appearance of *Species of Mind* (1999) by philosopher Colin Allen and behavioural ecologist Marc Bekoff. These authors took a less broadly comparative approach than Griffin had, balancing investigations of animals in the wild with animals in captivity and focusing on social play and vigilance (strategies to defend against predators) as markers of mental skills in non-human animals. Attempting to "read" cognitive abilities by testing animals with methodologies borrowed from tests applied to humans was self-defeating because of its glaring anthropocentricity. All too often investigators fail to use an experimental design appropriate for the world experience of the animal under study: "There is a need to develop and implement species-fair tests that tap the sensory and motor worlds of organisms belonging to different taxa" (Allen and Bekoff 1999).

A case in point is the question of how dogs discriminate self from others. The classic test to assess this is the mirror self-recognition test, which is dependent on the visual sense. In an exemplary model of thinking outside the box, ecologist Roberto Cazzolla Gatti (2016) noted that dogs, "as dolphins, show a high level of behavioural and cognitive complexity, but attempts to demonstrate self-recognition in these animals have been inconclusive because of difficulties in implementing adequate controls necessary to obtain robust evidence from the mirror test." Realizing that vision as sensory modality was inadequate in an animal whose prime connection to the world is through the sense of smell, Gatti developed a sniff test of urine samples from a group of four dogs who had lived together for at least seven years. The results warranted the conclusion that dogs do recognize their urine from that of others and display some evidence of self-consciousness.

The case for animal consciousness, far from continuing to tread water, has gained some momentum thanks to this new attention to the personality profile of the animal under investigation, which in turn colours the choice of experimental design. The further one strays from primates, whose alleged higher cognitive abilities should not be too surprising, given their relative kinship to humans, the more convincing the case for animal intelligence at large can be made. How far back can one stray? The consensus of neuroscientists points to cephalopods – squid, cuttlefish, and octopus. Observations on learning and memory in these animals date back to British zoologist J.Z. Young (discussed in another context in chapter 4). Experiments by Young (1966) and others suggest that the octopus's cognitive skills are superior to those of its molluscan cousin *Aplysia*. The mental range of cephalopods is unmatched among invertebrates: complex associative learning, manipulative skills, and possibly the use of rapid changes in skin colour and texture to communicate between individuals or with other species – what Peter Godfrey-Smith (2016) called "a kind of ongoing chromatic chatter."

If cephalopods developed a crude form of language code through their skin, what is one to make of the vocalization of parrots? Parrots can eerily imitate human speech not only thanks to their purported intelligence, but also because their throat's vocalization apparatus, unlike that of other birds, resembles in some respects that of humans. This was made clear by American neuroethologist Irene M. Pepperberg, who has studied the grey parrots's cognitive skills since the 1970s. In the book that synthesizes her work, *The Alex Studies: Cognitive and Communicative Abilities in Grey Parrots* (1999), Pepperberg made the point that "Alex" and other grey parrots that she studied are better amenable to incisive investigations of such skills than primates. At the onset she asked herself: "How much do these birds really understand? How much could these birds learn to understand? Given an appropriately enriched environment, might parrots turn out to be the great apes of the bird world?"

Pepperberg was able to answer these questions satisfactorily by her creative testing of these birds. Guiding her work was the view that intelligence is "an evolutionary outcome of the need not only for memory and flexibility, but also for choosing what to ignore as well as what to process." Her investigations suggested that Alex was doing just that. Alex, for instance, was able not only to discriminate between objects by their features, but more importantly to derive the very concept of sameness or difference between objects. From her results Pepperberg proposed "that the combination of intelligence and advanced communication skills may have arisen not only in primate or even mammalian lines, but also in avian lines, and that it directs not just learning but also what is appropriate to learn" (Pepperberg 1999).

It goes without saying that complex cognitive skills such as those displayed by cephalopods and parrots pose a challenge to the reductionist neuroscientist attempting to unravel the brain architecture and web of circuitry behind these skills. So far no such effort has surfaced, and it will be a long road to reach this understanding. But suffice it to say that the labour of cognitive ethologists like Pepperberg and many others, who see the bigger picture of animal minds in the natural world, have been able to do away with the view that animals are directed merely by instinct.

10

Learning How Chemicals
Rule Animal Moods

ᴄᴠ

Chemical signals connect most of life's processes,
including interorganismal relationships.
~ Jelle Atema (1995)

In the previous chapter there was a deliberate bias to depict what neurobiologists had learned about animal brains in terms of neurons and circuits,
looking at the nervous system as if it were, to borrow from the language of
electrical engineers, hardwired and all sparks. From a historical perspective
this depiction is largely accurate, in that neurophysiologists used to view
the nervous system as just that – hardwired. But, when it was discovered
that animal brains were as much about soups as about sparks, this view
gradually shifted. Chemicals are used for signal transmission from neuron
to neuron or neuron to muscle. Hormones as a class of chemicals, which
were already known to exert actions throughout animal bodies, were later
revealed to exist in the brain as neurohormones. And still later, other substances such as pheromones were found to mediate communication at another level, between individuals. This chapter follows these developments
and what they mean for our understanding of the way animal life is ruled
by chemical signals.

The story of the discovery of chemical neurotransmission has been told
admirably in two books: first by Jean-Claude Dupont in his *Histoire de la
neurotransmission* (1999), and shortly afterward by Elliot S. Valenstein in *The
War of the Soups and the Sparks* (2005). In addition, a Belgian physiologist

who was on the front line of early events in this story, Zénon-Marcel Bacq (1903–1983), provided his own personal account (Bacq 1975). The following is but a distillation of their contributions and, for lack of space, is selective in its treatment, emphasizing coverage of animals at large at the expense of biomedical aspects.

The notion of chemical neurotransmission began to take shape at the dawn of the twentieth century thanks to John Newport Langley (1852–1925), a product of Michael Foster's Cambridge School of Physiology who became a distinguished physiologist and pharmacologist and succeeded Foster as Cambridge's chair of physiology. Langley was at a loss to explain why the action of cat adrenal gland extracts in his experiments "should correspond so closely with that caused by stimulation of the sympathetic nerves" (Langley 1901). It behooved Langley's student Thomas Reston Elliott (1877–1961) to pursue this line of investigation. Elliott reasoned that the "marked functional relationship of the [adrenal glands] to the sympathetic nervous system harmonises with the morphological evidence that their medulla and the sympathetic ganglia have a common parentage. And the facts suggest that the sympathetic axons cannot excite the peripheral tissue except in the presence, and perhaps through the agency, of the adrenalin or its immediate precursor secreted by the sympathetic paraganglia." Elliott's experiments obliged him to conclude that "adrenalin might then be the chemical stimulant liberated on each occasion when the [nerve] impulse arrives at the periphery" (Elliott 1904).

If a transmitter substance is released at nerve endings upon excitation, as Elliott implied, then what happens next? How does the released transmitter act on the receiving cell at the synapse? The first tentative answer came from Elliott's mentor. Langley (1906) tried to make sense of drug actions, some of which may imitate or antagonize a neurotransmitter's action. He was particularly intrigued by the paralysing actions of curare – the poison from plant extracts used by South American natives on their arrowheads – and nicotine. His analysis of the drugs' actions led him to his receptor theory, proposing "the presence in [the receiving cell] of one or more substances (receptive substances) which are capable of receiving and transmitting stimuli, and capable of isolated paralysis, and also of a substance or substances concerned with the main function of the cell (contraction or secretion, or, in the case

John Newport Langley.
Image from Wellcome Collection.

of nerve cells, of discharging nerve impulses)." Only later was the theory validated and the receptors found to be specific and diverse proteins imbricated in the cell membrane – be it nerve, muscle, or other cell types – acting like keyholes to the specific transmitter key.

As for chemical transmitter release, it took another seventeen years after Elliott's hypothesis for the proof to materialize, in this case for a suspected

Otto Loewi.
Image from Wellcome Collection.

Walter Bradford Cannon.
Image from Wellcome Collection.

neurotransmitter other than adrenaline. Otto Loewi (1873–1961) was the German pharmacologist who delivered the decisive experimental demonstration. A medical physiologist by training, Loewi came relatively late in his career, at forty-eight, to the vexing problem of chemical neurotransmission, when he was chair of pharmacology at the University of Graz in Austria (Dale 1961). Using isolated frog hearts bathed in a physiological saline solution, Loewi showed that by stimulating the vagus nerve of one heart and collecting the post-stimulation fluid, he was able to produce the normal effect of vagus nerve stimulation – slowing and weakening of the heartbeat – with that fluid on another heart devoid of innervation (Loewi 1921). He called the substance in the fluid, likely released from nerve endings, *Vagusstoff*, later found to be the ubiquitous neurotransmitter acetylcholine (Loewi and Navratil 1926).

Meanwhile, efforts to follow up on Elliott's hunch about adrenaline as a neurotransmitter were going nowhere. The first encouraging step in this pursuit was the discovery of an unidentified substance, sympathin, produced by the action of nerves of the sympathetic nervous system on smooth (visceral) muscles (Cannon and Bacq 1931). This pursuit was very personal to Walter B. Cannon, whose concept of homeostasis we discussed in chapter 3. Another important concept introduced by Cannon was the fight-or-flight response, according to which emotions experienced by animals – pain, hunger, fear, rage – affect stomach and other internal functions via the mediation of hormonal (adrenal glands) and nervous (sympathetic nerves) paths (Cannon 1915). The concept of stress later developed by then McGill University endocrinologist Hans Selye (1936, 1938) evolved from Cannon's vision, even though Selye never acknowledged it.

Cannon and Bacq's 1931 report on sympathin betrayed the ambiguities of the age about chemical neuromediation, in that it was viewed through the lens of endocrinology. The title of their article speaks of "A hormone produced by sympathetic action on smooth muscle." Cannon's review article of the same year, titled "Recent studies on chemical mediations of nerve impulses," was published in the journal *Endocrinology*. The study of hormones was initiated around the middle of the nineteenth century (Medvei 1982), about the same period when bioelectricity took off under the impetus of Emil du Bois-Reymond (see chapter 2). The parallel evolution of these two fields encouraged the notion that soups (hormones) and sparks (nerve impulses) are separate compartments involved in the coordination of animal activities. As a result, investigators such as Cannon had a hard time teasing out neurotransmitters from hormones, and neurophysiologists could not imagine that nerve function has a chemical element.

Cannon and Bacq (1931) pointed out that sympathin shared many functional features with adrenaline, but did not go as far as to state they are the same substance. There followed misinterpreted investigations that embroiled the involved labs in controversies (Bacq 1935). Bacq himself proposed that sympathin was in fact a derivative of adrenaline, noradrenaline (Bacq 1934). The controversy was resolved when Loewi (1936) showed that in his pet model (frog) adrenaline is the neurotransmitter, and when the Swedish pharmacologist Ulf Svante von Euler (1905–1983) identified noradrenaline as the sympathetic transmitter in Cannon and Bacq's pet model, the cat (Euler 1946). For his neurotransmitter work, Von Euler shared the 1970 Nobel Prize in Physiology or Medicine.

What these controversies and resolutions brought about was the realization that neurotransmitters and their mechanisms may differ among animal groups. This new understanding laid the foundation of the field of comparative pharmacology. Early on, since it was assumed that all vertebrates functioned with acetylcholine and adrenaline/noradrenaline, comparative pharmacologists paid more attention to the bewildering diversity of invertebrates to test how ubiquitous these neurotransmitter systems are. Zénon Bacq (1947) provided the necessary leadership in his comparative account of the distribution and pharmacology of acetylcholine and adrenaline in invertebrates from protozoans to mollusks and arthropods. As early as 1934 Bacq, working from the Naples Zoological Station, published profusely on this topic. The evidence he obtained was disappointing; for example, except for leeches and echinoderms (sea cucumbers), the case for acetylcholine as a neurotransmitter proved unconvincing. From the moment researchers began thinking outside the box, the field widened: what if the scope of neurotransmitters extended far beyond acetylcholine and adrenaline?

Another substance related to monoamines besides adrenergic transmitters turned up in the 1930s (Whitaker-Azmitia 1999). In 1937 Vittorio Erspamer (1909–1999), an Italian pharmacologist, and his supervisor Maffo Vialli (1897–1983) at the University of Pavia (Mazzarello 2009), extracted from the rabbit gut wall a substance they called enteramine. By the early 1950s it was realized that enteramine was in fact serotonin, a substance identified in mammalian blood serum (Rapport, Green, and Page 1948). Soon a remarkable graduate student in the lab of Harvard physiologist John H. Welch (1901–2002) discovered that serotonin relaxed the catch muscle of mussels and might act as a neurotransmitter. Betty Mack Twarog (1927–2013) made her discovery in 1951, but, as Patricia Whitaker-Azmitia (1999) explained, "Although Dr. Twarog wrote the paper on these findings immediately, the paper was not actually published until two years later (Twarog 1954) because the *Journal of Cellular and Comparative Physiology* had not bothered to review a paper on an unknown neurotransmitter by an unknown author." In 1953 Twarog and Page detected serotonin in mammalian brains and this landmark discovery opened up studies of a neurotransmitter role for serotonin in the brain.

Among the many physiological roles of serotonin unveiled over the second half of the twentieth century, its role in modulating mood or the emotional make-up has stood out for its implications in disorders such as depression, anxiety, and schizophrenia. Similarly, serotonin was found to modulate behaviour in some invertebrates. In fact, serotonin is ubiquitous in the natural world, being present even in plants, and is now considered to be one of the most ancient neurotransmitters in the animal world. Its role in modulating behaviour is also widespread among invertebrates (Weiger 1997). A remarkable example discovered by Edward Kravitz's team at Harvard is how serotonin and octopamine – the substitute for adrenergic transmitters in many invertebrates – promote aggressive and submissive postures, respectively, in lobsters (Livingstone, Harris-Warrick, and Kravitz 1980). Antagonistic roles for these two biogenic amines were even traced back to nematode worms (Horvitz et al. 1982).

Other important neurotransmitters discovered in the 1950s are the amino acids glutamate and gamma-aminobutyric acid (GABA). It made more sense to interpret the large amounts of glutamate in the mammalian brain as the result of its role in energy metabolism or as a protein constituent, but its role as an excitatory neurotransmitter gradually emerged in the 1950s and 1960s (Watkins and Jane 2006). It started with the effects of injecting glutamate in the motor cortex of dogs (Hayashi 1954) and progressed to actions on specific nerve cells (Curtis, Phyllis, and Watkins 1960). Meanwhile glutamate turned up as a neuromuscular transmitter in crustaceans (Harreveld and Mendelson 1959; Takeuchi and Takeuchi 1964).

So far in this story, each discovered chemical transmitter was viewed as being associated with excitatory nerve endings, and although inhibitory nerves were known in crayfish (see chapter 9), the quest for an inhibitory transmitter was ongoing. An Austrian neurobiologist, Ernst Florey (1927–1997), working in Cornelis Wiersma's lab at Caltech, had extracted from the brain and spinal cord of mammals a neuroactive substance (Jasper 1984). Florey tested his extract on a crayfish muscle and found that it inhibited neurons sensitive to muscle stretch; hence the name Factor I (for inhibition) that he gave the substance. Florey travelled next to the Montreal Neurological Institute of McGill University to work with Kenneth Allan Caldwell Elliott (1903–1986), a South Africa–born neurochemist. Elliott and Florey purified the extract and, after many twists and turns, identified it as the amino acid GABA (Bazemore, Elliott, and Florey 1956). It later became evident that the excitatory/inhibitory tandem of glutamate and GABA predominates in the central nervous system of vertebrates and in the peripheral (neuromuscular) nervous system of higher invertebrates.

"Synaptic transmission," Florey (1984) came to realize, "turns out to be a multifaceted and complex process. Certainly it involves chemical messengers, but messengers are released also from non-synaptic sites, and the release, whether it occurs at synapses or elsewhere, is not necessarily coupled with nerve impulses; hence the 'transmitter' function cannot be understood in the sense of 'transmission of nerve impulses.'" How did neuroscientists reach this conclusion?

∼

We moved from the paradigm of neurotransmitters taking care of the nervous system and glandular hormones working separately from the blood stream, to a relativistic paradigm in which there is room for more than one transmitter in the same nerve ending and for neurons that actually secrete hormones.

This turn of events came about, at least initially, from the rise of comparative endocrinology. Starting in the 1850s, discoveries of mammalian glands producing secretions that enter the bloodstream and act on distant organs had shaped the field of endocrinology (Medvei 1982). Interest in what happens to non-mammalian animals endocrine-wise manifested itself early in the twentieth century. As early as 1910 Bernardo A. Houssay (1887–1971), the

Argentine endocrinologist honoured with the 1947 Nobel Prize, made a number of discoveries on the pituitary gland using the toad as model animal. He selected the toad for "its abundance and cheapness, its resistance to trauma, the facility of operative techniques, the great number and clarity of the symptoms of pituitary insufficiency, the rapidity and intensity of the reaction to implantation of any of the lobes of the pituitary, and the possibility of making experiments and obtaining proofs more easily and in larger numbers than with any other animal" (Houssay 1936). Interest in invertebrate endocrinology was also manifested early, as collated by the German zoologist Gottfried Koller (1929). These early works relied on the effects of gland extracts and gland surgical removal on animal functions, as the chemical identification of the active hormones awaited methodological innovations.

A revolutionary outcome of comparative endocrinology was the slow emergence of the concept of neurosecretion and the field of neuroendocrinology. This development began with discoveries by zoologists Carl C. Speidel (1893–1982) and Ernst Scharrer (1905–1965). Speidel, an anatomist at the University of Virginia, described in 1919 unusually large glandular cells in the tail part of the fish spinal cord. In a prescient statement from 1922 he noted that "the structure of the cells from a comparative standpoint suggests a series of transition stages from primitive nerve tissue to glandular tissue." It was later realized that these neurons secreting blood-borne hormones formed a pituitary of the tail, so to speak, called the urophysis, involved in osmoregulation (Fridberg and Bern 1968). Scharrer was a doctoral student in Karl von Frisch's lab in Munich (see chapter 4) when he discovered neurosecretory cells in the hypothalamus of a minnow (Scharrer 1928). He inferred their neuroendocrine role from the proximity of small blood vessels and he suspected, rightly, that these cells communicate with the nearby pituitary gland.

At the time Scharrer published his article, there was a conceptual confusion about the meaning of neurosecretion. Loewi is recognized as providing the first experimental evidence of neurotransmitter activity, but his seminal 1921 paper couched his finding in terms of a "neurohumoral" phenomenon, as if the neurotransmitter, once released from nerve endings, diffused far away as a hormone would. This ambiguity was echoed by Cannon, who seemed to handle the wordings "sympathetic transmitter" and "hormone" interchangeably. Their inability to fathom the physical arrangement of nerve synapses, technically excusable at the time, meant that neurotransmission was misconstrued for something that Scharrer actually got right: neurohormonal secretion.

Sharrer met his future wife in Frisch's lab, where she was also pursuing her doctoral degree. Ernst and Berta Scharrer became partners in research as in life once Ernst was appointed director of the Neurological Institute of the University of Frankfurt in 1934 (Purpura 1998). While Ernst concentrated on gathering evidence of neurosecretion across vertebrate classes, Berta did the same for invertebrates, from annelid worms to crustaceans and insects (Scharrer and Scharrer 1937). The Scharrers' moral opposition to the Nazi regime prompted their move to the United States in 1937, where they continued to make their case for the concept of neurosecretion in the face of resistance or skepticism from many colleagues. Over the years the concept gained ground, and it is now universally accepted that across the animal world some neurons can secrete hormones that act indirectly through their control of endocrine glands (pituitary in vertebrates, prothoracic glands in arthopods, for example) or directly via the bloodstream on more distant functional targets (osmoregulation, growth, and so forth).

Comparative endocrinology kept thriving as a field along with neuroendocrinology on the side. The discoverer of the hormone prolactin, Oscar Riddle (1877–1968), presented a dynamic snapshot of the field in his 1935 article "Contemplating the Hormones": "Present pursuit of the hormones," he asserted, "is at such a pace that we may well be persuaded that endocrinologists are just now in a bigger hurry than any other group of investigators in the world. In this we are right: never before were so many choice secrets of chemical regulation spread just beyond our finger tips." Early on, endocrinologists were finding out how widely the identified hormones cross-react among vertebrates. Riddle's own prolactin, which he identified in pigeons (Riddle et al. 1933), acts as a coordinator of maternal behaviour, such as lactation in mammals and nesting in birds, and also as an osmoregulator in freshwater fishes, preventing salt loss and water intake. Similarly, the thyroid hormone thyroxine, chemically identified by the American Nobel Prize-winner Edward C. Kendall (Kendall and Osterberg 1919), is present not only in the thyroid gland of all vertebrates, but also in the ascidian endostyle, considered to be the evolutionary precursor of the thyroid gland (Gorbman 1941).

Bertil Hanström in 1925.
From photo archives of the Marine Biological Laboratory, Woods Hole.

Endocrinologists studying invertebrates had a more arduous task because invertebrates are inordinately diverse compared to the relatively monolithic vertebrates. The comprehensive surveys of hormones and endocrine functions of invertebrates by Gottfried Koller (1937) and the Swede Bertil Hanström (1939) gave pause and forced endocrinologists to challenge the classical definition of hormones. Hanström laid out the scope of the challenge:

The old definition of a hormone as a substance which is secreted into the blood by a ductless gland and exerts a specific physiological effect at another place in the body (even if present in very small quantities) is no longer satisfactory. Especially in invertebrates, in which there is no relation between the many known instances of physiological processes which must be classified as hormonal and the relatively few structures identified as true [endocrine] organs analogous with those of vertebrates, the old definition cannot be used. Furthermore, there are many invertebrates which do not possess blood and in which the hormones, if they exist, must be transported by means of diffusion through the protoplasm. We also nowadays know several substances whose action must be hormonal, but which are certainly not produced by [endocrine] glands but in portions of organs whose chief function is quite other than hormonal.

Koller and Hanström came up with a definition simple enough to account for all the idiosyncrasies of the hormonal world: hormones are organic substances produced by the organism for its own use, active in very small amounts and endowed with a specific regulatory function. The way hormones played out, as Koller, Hanström, and future endocrinologists ascertained, ran the gamut of levels: within a cell (intracrine), on cells close to the secretory source (paracrine), on cells and tissues distant from a secretory source through diffusion or the bloodstream (endocrine). As the field progressed, the chain of command of endocrine functions in higher invertebrates and in vertebrates emerged ever so sharply. Looking at mammals, fish, crayfish, or insects brought analogous endocrine paths into focus: a cascade from the brain to neuroendocrine centres to endocrine organs to regulated functions.

These developments climaxed in the 1960s as they were reviewed in a two-book series, *Comparative Endocrinology* (1963), edited by Ulf S. von Euler. The first volume dealt with the hypothalamus-pituitary neuroendocrine axis in vertebrates, pituitary hormones such as those controlling changes in skin coloration in frogs, thyroid and pancreatic hormones, reproductive hormones, and so on. In the second volume invertebrate neurosecretions and hormones involved in reproduction and growth (moulting) are discussed as well as locally acting hormones. One of the chapters was authored by Aubrey Gorbman (1914–2003), who was to leave an indelible mark on the field.

Gorbman was born in Detroit to Russian immigrants and studied biology at Wayne State University where his master's research dealt with the evolution of thyroid function in vertebrate ancestors and early vertebrates (Gorbman and Creaser 1942). He completed a PhD in zoology at Berkeley. His first professorial post was at Barnard College of Columbia University, where he pursued his interest in the vertebrate thyroid and iodine uptake in invertebrates. Once he took a professorship of zoology at the University of Washington in Seattle, he settled there until his death. It is said that Gorbman embraced comparative endocrinology because he was allergic to mammals (Smith 2003). The gain for comparative endocrinology was immense, as he emerged as a leader of the field during his tenure at Barnard and in Seattle.

In the early 1960s Gorbman's leadership materialized in the foundation of the journal *General and Comparative Endocrinology* and the publication – co-authored with his close friend and colleague Howard A. Bern (1920–2012) – of the seminal *A Textbook of Comparative Endocrinology* (1962). Gorbman (1993) recalled how the journal was founded:

It derived from a discussion in 1961 between Choh Hao Li, the pituitary hormone biochemist, and Kurt Jacoby, a founder of a then-young publishing house in New York that specialized in titles that had relatively limited sales potential and high risk in returning to the publisher the cost of their production. Jacoby and Walter J. Johnson had brought Academic Press to New York from Germany in the early 1940s. Jacoby was interested in starting several journals and he was persuaded by Li that General and Comparative Endocrinology would fit the pattern that Academic Press had established for itself at the time.

Gorbman was chosen as the American editor of the journal – and remained in this post for thirty-one years – with the British endocrinologist (and concert organist) Ernest J.W. Barrington (1909–1985) as European editor. Although the journal was not the only venue available to zoologists with an endocrinological bent, it remains to this day a major organ for the field. Its mission converged with the goal of Gorbman and Bern's textbook to scout for "trends in the hormonal control of adaptation and evolution," and to "attempt to integrate the latest findings in comparative endocrinology, neuroendocrinology, and the cellular level of hormonal action from a non-medical approach" (Freeman 1963).

An important figure in comparative endocrinology who interacted with Gorbman was the French zoologist Maurice Fontaine (1904–2009, see also chapter 7). Fontaine shared with Gorbman a keen interest in thyroid function in the lowest of the vertebrates (lampreys) that led to a joint publication (Fontaine et al. 1952). The Frenchman was intrigued by the reproductive migrations of eels and salmon, and he spent a great part of his career exploring them. He did not let the crude state of endocrinological affairs in his time inhibit his pursuit. Already in the 1930s he wondered if something akin to the mammalian pituitary hormone triggering sexual maturation is involved in the development of the sexual organs of migrating eels. So he took urine samples of pregnant women, known then to contain the pituitary reproductive hormonal factor, and injected them into immature eels kept in seawater, their environment when they migrate to their breeding grounds. He found that the human urine samples caused not only the growth of the sexual organs but also the appearance of secondary sexual characteristics (Fontaine 1936). Fortunately, the future of comparative endocrinology did not depend on such methods to make advances.

Comparative endocrinology, much as its paramedical counterpart, was dominated by men. Yet a woman, British zoologist Penelope M. Jenkin (1902–1994), had produced a book, *Animal Hormones: A Comparative Study*, in 1962. In contrast to Gorbman and Bern's textbook of the same year, Jenkin's book concentrated on the action of hormones in vertebrates and invertebrates "rather than on their sources or on their phyletic distribution." To make sense of the bewildering diversity of hormones and their pervasive actions, she separated kinetic from metabolic and morphogenetic hormones. By kinetic she meant hormones acting on muscles, pigmented cells (chromatophores), digestive, milk-secreting and skin glands, and even on endocrine glands. Metabolic hormones she classed as those that affect respiration, carbohydrate and protein metabolism, and water and mineral balance. Morphogenetic hormones are especially involved in growth and metamorphosis.

Another woman of significant impact in comparative endocrinology was Grace Evelyn Pickford (1902–1986). Born in Bournemouth, England, she studied zoology at Newnham, a Cambridge women's college (Slack 2003). In 1925 she married her fellow Cambridge student G. Evelyn Hutchinson, who later counted among the top ecologists of his generation. From 1925 to 1927 the pair spent time in South Africa, where Grace did extensive research on earthworms. Hutchinson having secured an instructor job at Yale in 1928, the couple moved to New Haven. As Slack explained, as a condition of his acceptance of the instructorship, "Hutchinson had asked for and been assured of research space in the zoology building for his wife." At Yale she completed a PhD out of the South African earthworm material. "Throughout the thirties, forties and most of the fifties," Slack added, "she was a Fellow, a Research Assistant and, from 1946, a Research Associate at the Bingham Oceanographic Laboratory at Yale."

Pickford's passion for all things zoological, and her determination to build a research career despite the obstacles thrown in her path on account of being a woman, led to her promotion at Yale to associate (1959) and full professor (1969, a year before her forced retirement). It took a meandering path, along which she participated in oceanographic expeditions and became an expert on squids, before she started work in endocrinology after World War II. Her experimental model was the estuarine killifish (*Fundulus*), on which she began publishing in 1948. As her close colleague James W. Atz (1986) put it, "In her hands, *Fundulus heteroclitus* has illuminated comparative endocrinology." Pickford's lab at Yale became a centre of attraction for fish endocrinology, culminating in a book co-authored with Atz, *The Physiology of the Pituitary Gland of Fishes* (1957). After retiring from Yale in 1970, Grace took the post of "Distinguished Scientist in Residence" at an undergraduate college in Ohio, Hiram College. There she pursued her research, relying for the most part on the assistance of undergraduate students, and she kept publishing until 1984, two years before her death.

Since the days of Grace Pickford and her contemporaries, the field of comparative endocrinology has expanded and taken a molecular turn. From among her circle of disciples, fish endocrinologists such as Canadian Richard E. Peter (1943–2007), for one, helped develop the field of fish neuroendocrinology. Neuropeptides – often made of short chains of amino acids – became a mainstay of neuroendocrinological research not only in fish and other vertebrates but also in invertebrates. David A. Price and Michael J. Greenberg at the University of Florida identified in 1977, for example, a small cardioactive peptide from a mollusk, dubbed FMRFamide, which turned out to belong to an important family of neuropeptides widespread across invertebrate phyla and present even in mammals. As more hormones and their receptors were discovered and their genes characterized, their incredible functional diversity was increasingly appreciated. In an article entitled "Comparative endocrinology in the 21st century" (2009), Robert J. Denver and a host of modern practitioners offered this assessment of the field:

> Comparative endocrinologists work at the cutting edge of the life sciences. They identify new hormones, hormone receptors and mechanisms of hormone action applicable to diverse species, including humans; study the impact of habitat destruction, pollution, and climatic change on populations of organisms; establish novel model systems for studying hormones and their functions; and develop new genetic strains and husbandry practices for efficient production of animal protein. While the model system approach has dominated biomedical research in recent years, and has provided extraordinary insight into many basic cellular and molecular processes, this approach is limited to investigating a small minority of organisms ... A major challenge for life scientists in the 21st century is to understand how a changing environment impacts all life on earth ... Comparative endocrinologists have a key role to play in these efforts.

Hormones can affect brain circuits involved in the display of sexual or other behaviour, or they may influence gene expressions involved in coordinating mood changes. This subfield of endocrinology has grown over the years, leading even to a journal devoted to the topic, *Hormones and Behavior*, founded in 1969. The environmental control of hormone-behaviour relationships has also come into sharp focus. In this respect, hormones are not the only players.

~

A different class of chemicals operating at a different level emerged in the early twentieth century: pheromones. They were first regarded as "ectohormones" in the sense that they worked like air-borne or water-borne hormones released from an individual and detected as a meaningful signal by another individual.

The first researcher to have stumbled on this concept was the famous French entomologist Jean-Henri Fabre (1823–1915). He watched emperor moths (*Saturnia*) find their mate from far away at night and he had to discard vision as guide (Fabre 1913). He asked himself: "What sense is it that informs this great butterfly of the whereabouts of his mate, and leads him wandering through the night? What organ does this sense affect?" It took numerous experiments for Fabre to arrive at the conclusion that smell, through the moth's olfactory organs, the antennae, was responsible. It was still hard to believe for the anthropocentric investigator familiar with the capabilities if the human nose, so he asked: "Are there effluvia analogous to what we call odour: effluvia of extreme subtlety, absolutely imperceptible to us, yet capable of stimulating a sense-organ far more sensitive than our own?" After eliminating memory of location by decisive experiments, he answered his own question in the affirmative.

Chemical communication in the environment can take other forms, as zoologist and ethologist Karl von Frisch (discussed in chapter 4 as a pioneer of German comparative physiology) found out. In a paper ambiguously titled (my translation) *On the Psychology of Schooling Fish* (1938), he noticed that an injured minnow can elicit fright reactions in the school of minnows to which it belongs. Like Fabre, Frisch conducted numerous experiments leading him to conclude that the fish's injured skin releases what he called a *Schreckstoff*, or alarm substance, that alerts specifically the fishes' congeners of a potential danger looming. The substance, Frisch determined, was detected by smell, not taste.

The third player in this story was the German chemist Adolf Butenandt (1903–1995). His accomplishments deserve special attention. Butenandt was a precocious youngster who had already proved his mettle in chemistry in high school (Akhtar and Akhtar 1998). He studied chemistry and biology at Marburg University and earned his PhD in organic chemistry in 1927 at

Göttingen University. While pursuing postdoctoral studies in Göttingen, Butenandt undertook to purify and determine the chemical structure of the female sex hormone. To accomplish this he started where Maurice Fontaine had left off, with urine from pregnant women; a pharmaceutical company "supplied Butenandt with a dark brown syrup, extracted from a large volume of urine from pregnant women" (Akhtar and Akhtar 1998). Remarkably, at a young age between 1928 and 1934 Butenandt was able to identify not only the female sex hormone (oestrogen), but also the sex hormones progesterone and testosterone. And after being appointed director of the Kaiser Wilhelm Institute of Biochemistry in 1935, he continued to accumulate chemical discoveries. No wonder he was the recipient, at only thirty-six, of the 1939 Nobel Prize in Chemistry.

Butenandt did more than sit on his laurels. After World War II he moved to Tübingen University where in 1954 he isolated and identified the hormone that controls moulting in insects (Butenandt 1959). It was called ecdysone and it turned out to be a steroid like the mammalian sex hormones. But the relevant discovery here is the first chemical identification of a pheromone in an insect. Peter Karlson (1995), who spent time in Butenandt's lab, gives a nutshell account of the scale of the research:

The silkworm's sex attractant is produced in small glands sitting at the tip of the female abdomen, and is released to lure the male moths for copulation. Isolating it was again a fight for enough starting material. The great campaign of 1953 yielded 200,000 sex glands, but even that was not enough. Butenandt and his co-worker Hecker had to order 500,000 glands from Japan, and from this material the attractant was extracted and its structure determined. This was the first insect pheromone to be isolated.

The substance was named bombykol (from *Bombyx*, the scientific name for the silkworm). In a paper reporting the discovery, Butenandt (1959) uses the word pheromone for the first time. This designation was actually proposed by Karlson and Lüscher in 1959. They explained that the newly coined word "is derived from the Greek *pherein*, to transfer; *hormōn*, to excite." Butenandt was initially reluctant to embrace the word, preferring the old form ectohormone, but he was persuaded by Karlson to adopt it. Also in

1959 Karlson and Butenandt were already reviewing what was known of the workings of pheromones, even though, with the exception of bombykol, their chemical structure was still unknown. They identified sex attraction, warning signals, and territorial demarcation (urine marking) as their purview. Because pheromones are inextricably linked with social behaviour, it is no surprise that they are particularly active in social insects such as bees and termites.

As the chemical identity of more pheromones was known, it became apparent that their molecular weight tends to be quite small so they can travel fast, especially as alarm signals, and far, as sex attractants (Wyatt 2009; Steiger 2011). In the course of evolution, researchers argued, the organ of smell developed not only specificities for particular molecular scents but, as importantly, extremely high sensitivities to pheromones that arrive much diluted at their target. Pheromones can be detected at nanomolar or even picomolar concentrations by olfactory cell receptors, which means that their performance is a thousand times or a million times better than that of hormone receptors.

Another form of chemical communication in the environment is olfactory imprinting. Sensory imprinting was first described in geese by Karl Lorenz (1935), in a paper explaining how freshly hatched goslings imprint on the first moving object they see and assume it is their mother. Imprinting associated with the sense of smell, on the other hand, left an indelible mark in popular science in relation to the homing behaviour of salmon. An important player in this story is the freshwater ecologist Arthur David Hasler (1908–2001). As Gene Likens (2002) points out in a memoir of his life, Hasler's "work on the mechanisms whereby salmon find their way back from ocean feeding areas to home streams for spawning, for which he was best known, was not only brilliant and innovative but also provided a framework for management of these important fisheries throughout the world."

Hasler was born in Utah to Mormon parents and he studied zoology at Brigham Young University in Salt Lake City (Likens 2002), In 1937 he obtained his PhD at the University of Wisconsin in Madison, where he was appointed assistant professor in 1941. After serving in World War II, he returned to Madison to resume his tenured professorship until his retirement in 1978. To the question of how he became involved in salmon homing behaviour Hasler answered retrospectively (Hasler and Scholz 1983) by asking: "How does a scientist go about the task of pushing back the curtains of the unknown? Certainly the romance of tackling the mysteries of nature provides the motivation, for who would not be inspired by the remarkable life history of this romantic beast, the salmon." Hasler eloquently explained the salmon's predicament:

> After living in the Pacific Ocean for several years, salmon swim thousands of kilometers back to the stream of their birth to spawn. I have always been fascinated by the homing migration of salmon. No one who has seen a 20-kilogram salmon fling itself into the air repeatedly until it is exhausted in a vain effort to surmount a waterfall can fail to marvel at the strength of the instinct that draws the salmon upriver to the stream where it was born. But how does it find its way back?

Hasler was inspired by two predecessors who have appeared earlier in our story: "I was puzzling over this problem during a family vacation in 1946. Inspired by the work of the great German Nobel Laureates, Karl von Frisch and Konrad Lorenz, I had been conducting research with my graduate student Theodore Walker, since 1945, on the ability of fishes to discriminate odors emanating from aquatic plants." In that paper about minnows, Walker and Hasler (1949) came to this conclusion:

> Results of our studies lend support to the view that aquatic plants may well play an important role in the life of a fish. They may serve as signposts to guide fish into feeding grounds, since many fishes commonly feed in turbid water, at dusk, at dawn, and at night, when visibility is poor. Moreover, the odors of aquatic plants may serve as attractants to immature fishes to prevent them from straying from cover. In addition, other natural odors may direct migratory fishes in locating their homing areas.

The latter statement impelled Hasler to ask whether salmon can detect and discriminate among stream odours. As the answer turned out to be affirmative (Hasler and Wisby 1951), Hasler proposed "that the nature of the

guiding odor must be such that it have meaning only for those salmon con-
ditioned to it during their freshwater sojourn. Any substance which was
merely a general attractant could not guide salmon to their home tributary."
From this realization to the concept of imprinting, Hasler had but a small
step to climb, as he recalled (Hasler and Stolz 1983):

> The connection caused me to formulate the hypothesis that each stream
> contains a particular bouquet of fragrances to which salmon become
> imprinted before emigrating to the ocean, and which they subsequently
> use as a cue for identifying their natal tributary upon their return from
> the sea. I envisioned that the soil and vegetation of each drainage basin
> would impart a distinctive odor to the water, thereby providing the
> salmon with a unique cue for homing. Later, I formalized this hypoth-
> esis in collaboration with my student, Warren Wisby, in 1951.

Decisive experimental evidence for imprinting came many years later
(Cooper et al. 1976). Another animal that attracted plenty of attention with
regard to olfactory imprinting is the green sea turtle. Archie Carr (1909–
1987), a famous American reptile specialist, proposed in 1967 that green turtles use their keen sense of smell to imprint the beach of their birth, to which
they return from the feeding grounds. But Carr and others were unable to
provide convincing evidence. The concensus today is that these turtles use
geomagnetic orientation to navigate their oceanic way to the approximate
coastal area, and olfactory imprinting to locate the precise nesting ground.

In the continuum of chemical communication categories, one cannot ignore chemical defences. It soon became apparent early on that this mode of
defence evolved primarily in animals not quick enough to dart away from
a danger zone. If animals can release chemical attractants, they can also
broadcast chemical repellents to advertise to a potential predator that they
are unpalatable. One is reminded of the extreme example of the foul-
smelling skunk, but researchers have emphasized how widespread the phenomenon is in the animal world. As early as 1874 Belgian entomologist Ernest
Candèze (1827–1898) called attention to the use of repellents by vulnerable
insect larvae, and ever since then, zoologists have asked how the phenomenon and its mechanisms evolved in crustaceans and insects (Eisner
and Meinwald 1966).

Sometimes offence is the best defence, as witnessed in animals that possess glands producing toxic venoms designed to maim or kill their victim or attacker. Although toxins produced by animals had been known since antiquity, and physiological mechanisms of poison toxicity had been elucidated, as Dietland Müller-Schwarze points out in his book *Chemical Ecology of Vertebrates* (2006), "ecologists have investigated why animals and plants have poisons and venoms in the first place only since the 1950s." When examining the hundreds of fish species that are notoriously toxic, none beats the pufferfish. Its toxic notoriety was already known to the ancient Egyptians, and Captain James Cook came close to death eating the fish when visiting New Caledonia (Müller-Schwarze, 2006). It was revealed that pufferfish obtain tetrodotoxin, a muscle-paralyzing toxin, from grazing on algae and bacteria, and store it in various organs except muscles (Yasumoto and Murata 1993). Pufferfish use the poison for protection, but other animals, such as snakes, can use their venom to paralyse a prey, thus making it easier to grasp and swallow.

∼

In this chapter we have travelled through what we viewed as different levels of chemical communication, from the most localized (neurotransmission), to body range (neurohormones and glandular hormones), and finally to environmental chemical messengers. But in reality the categories of chemical messengers are not sealed tight. Researchers have learned over the years that some neurotransmitter substances such as serotonin can also act as neurohormone; that some brain neurotransmitters such as acetylcholine act on local receptors right across the synaptic gap, while others like noradrenaline diffuse further away to receptors at a distance from the release point (Descarries and Umbriaco 1995); and that two different neurotransmitters can coexist in the same nerve ending, and so on.

These relativistic strategies of chemical messengers have raised questions about how they came to be. American endocrinologist Jesse Roth is among those who have searched for answers. Roth was working on mammalian insulin in 1980 when he and his team at the Diabetes Branch of the National Institutes of Health found an insulin-like hormone in single-cell organisms such as protozoans and molds (LeRoith et al. 1980). They saw evolutionary implications in their discovery:

The argument whether the endocrine system is a descendant of the nervous system (or vice versa) might really be turned around to say that both arose from a common simple precursor. This would help to explain many previously unexplained findings, including the ubiquitous presence of insulin and chorionic gonadotropic hormone in all mammalian tissues, the extraordinary overlap of chemical messengers of the gut and the brain and the widespread similarities between the endocrine and the nervous systems.

Roth's team (Roth et al. 1986) reinforced this stance by comparing the apparent compartmentations "that separate the endocrine system and the nervous system, vertebrates and nonvertebrates, multicellular and unicellular," to the Maginot Line breached by Hitler's armies in 1940, which, although they "may appear to be formidable in theory, may provide no resistance when tested by data." Later in his career Roth paid attention to one functional implication of his theory: the gut-brain axis which has gained considerable currency today. His team (Hsiao et al. 2008) emphasized the role of the gut microbiome in generating signals to the brain for metabolic regulation. Such studies, they point out, "are leading to the recognition that the communities of microbes in the gut function as an 'organ' with many previously unappreciated metabolic, immunologic, and endocrine-like actions that influence human health. The true nature of this organ is rapidly being charted. What previously was considered a minor player in the sideshow is now approaching status as a star in the center ring."

These findings could not better illustrate the broad spectrum and interrelational nature of chemical communication processes in the animal world.

Epilogue

~

The place of the field of animal physiology in the early years of the twenty-first century has been examined from various angles by several practitioners. In 1998 Charlotte Mangum and Peter Hochachka offered their assessment of the new directions toward which the field should steer. "As we approach the end of the first century of comparative physiology and biochemistry," they wrote apprehensively, "a period of considerable growth and excitement, we find ourselves uncertain of our identity." Studying how "physiological systems work and how different kinds of animals are adapted to different kinds of environments" should still be a priority because some of the questions that physiologists have asked could not be answered with the technological tools available to them. The tools of molecular biology are now assisting in this regard.

Bringing fresh approaches to the question of how animals work is all very well, but what about the vexing problem of how animal functions evolve? In this book, evolutionary physiology has received scant attention. The reason can be found in Mangum and Hochachka's essay: "Early attempts to employ the evolutionary approach were not only few in number, they were unsatisfying in outcome because neither phylogenetic nor mechanistic/adaptational knowledge was adequate to serve as a firm foundation." Now that the phylogenetic relationships between animal groups are better understood and the tools of the comparative method – statistical modelling of functional trait distancing between species (Garland et al. 2005) – and molecular genetics can be harnessed, great strides should be expected in our understanding of the evolution of physiological mechanisms.

Another fresh approach linked to evolution that shows promise is developmental physiology. How do animal functions develop in the differentiating

and growing body? Warren Burggren and Stephen Warburton, animal physiologists whose research careers are largely devoted to this topic, produced
a review essay (2005) that defined such a research program. They stress the
interdisciplinary nature of the subfield: "Comparative developmental physiology spans genomics to physiological ecology and evolution." And they
show "how developing physiological systems are directed by genes yet respond to environment and how these characteristics both constrain and enable evolution of physiological characters."

These are new ways of addressing old challenges of comparative animal
physiology. A new challenge identified by Donald Mykles and his colleagues
(2010) revolves around the theme of integration; a function is not a stand-
alone entity independent of other body parts, species, or time. They identify
three grades of functional integration: "vertical integration of physiological
processes across organizational levels within organisms, horizontal integration of physiological processes across organisms within ecosystems, and
temporal integration of physiological processes during evolutionary change."
To meet this challenge physiologists will need to bring together a massive
amount of data to detect integrative patterns, to develop genetic model organisms such as the currently popular zebrafish, and to promote interactions
between practioners of related disciplines such as comparative physiologists,
evolutionary biologists, and geneticists (Mykles et al. 2010).

On the topic of interdisciplinarity one should mention the potential benefits of the discoveries of comparative physiology for clinical medicine. Clin-
ician Michael A. Singer addressed this in his book *Comparative Physiology,
Natural Animal Models and Clinical Medicine* (2007). In the book's introduction Singer gave a striking anecdotal example of what he is after:

> It is a winter day and Mr. Jones, a 45-year-old man suffering from
> chronic renal failure has just arrived at his regional dialysis center. He
> comes here three times a week, each time for four hours, to be con-
> nected to a hemodialysis machine. These treatments are necessary for
> Mr. Jones to stay alive. During each 4-hour treatment his entire blood
> volume will pass through the artificial kidney machine about 14 times
> for purification. On this same day, many miles away, an American black
> bear slumbers in its wintery cave. The bear will remain there dormant
> for up to five months during which time this animal will not eat, drink,

defecate or urinate. Although dormant, the bear still has an active metabolic rate about 50% of normal. Yet despite having no urine output for this prolonged period of time, the bear will not suffer any of the manifestations of renal failure experienced by Mr. Jones. How has the bear's metabolic machinery adapted to such a prolonged state of functional renal failure? Can we learn new approaches for the prevention and/or treatment of chronic renal failure from such a natural animal model?

He gives other examples in which birds and fish have found physiological adaptations that can provide insights into diseases such as diabetes and atherosclerosis. But other fields besides medicine can find inspiration from animal physiological adaptations. Biomimicry – innovation inspired by nature – is such a field. Of course biomimicry looms larger than physiological adaptations alone; Janine M. Benyus, in her book *Biomimicry* (1997), shows, for instance, how we can learn to weave fibres from watching spiders. Biomimicry, according to Kevin Passino (2005), can even assist computer engineering and automation by teaching how nervous systems work.

Finally, an important aspect of comparative animal physiology of great relevance to society today is environmental physiology, a term that encompasses ecophysiology or physiological ecology. In their textbook *Environmental Physiology of Animals* (2005), Pat Willmer and colleagues gave a nuanced description of what awaits the physiologist with an environmental interest:

Environmental adaptation is a complicated business, integrating all aspects of animal biology. It requires an understanding of animal design and animal physiology above all, but this must be put in context with a detailed understanding of the environment (measured on a suitable temporal and spatial scale), and with an appreciation of ecological and evolutionary mechanisms ... Equally there is a need to look beyond the confines of traditional isolated physiological "systems" (circulation, excretion, respiration, etc.) and to see the whole picture of what is needed in order to live in a particular environment: the physiological needs of course, but also the mechanical, sensory, reproductive, and life-history adaptations that together make up a successful fully functional animal.

In chapter 8 we got a taste of what animals can tolerate in extreme environments; their physiological adaptations are sometimes mind-boggling. But human activities, by encroaching on ecosystems, have unfortunately created an extreme environment of their own to which animals are at a loss to adapt. Witness lakes deprived of their oxygen and the death threat that deprivation poses to resident fishes. Witness chemicals dumped in water tables that bear too much resemblance to hormones and therefore can bind to hormonal receptors and disrupt normal endocrine functions. The World Health Organization and the United Nations Environment Programme have jointly issued a document (Bergman et al. 2013) in which horror stories about endocrine disruptors raise serious concerns for the health and even survival of humans as well as wildlife.

It is unfortunate that animal physiology, at this stage of its historical development, is increasingly called upon to sound the alarm about animals made dysfunctional by a toxic environment. Let us hope that this trend can be reversed and that we can continue to appreciate the beauty of animals harmoniously at work.

References

∽

Adler-Kastner, Liselotte. 2000. "Letter to the Editor: Some interesting historical details." *British Medical Journal* 320: 506.

Adrian, Edgar D. 1925. "The impulses produced by sensory nerve endings." *Journal of Physiology* 61: 49–72.

Akhtar, Muhammad, and Monika E. Akhtar. 1998. "Adolf Friedrich Johann Butenandt 24 March 1903–18 January 1995." *Biographical Memoirs of the Royal Society* 44: 79–92.

Albury, William R. 1974. "Physiological explanation in Magendie's manifesto of 1809." *Bulletin of the History of Medicine* 48: 90–9.

Ali, Heba A., Hin-Kiu Mok, and Michael L. Fine. 2016. Development and sexual dimorphism of the sonic system in deep sea neobythitine fishes: the upper continental slope." *Deep-Sea Research* Part I, 115: 293–308.

Allen, Colin, and Marc Bekoff. 1999. *Species of Mind: The Philosophy and Biology of Cognitive Ethology*. Cambridge: Massachusetts Institute of Technology Press.

Aminoff, Michael J. 2011. *Brown-Séquard: An Improbable Genius Who Transformed Medicine*. New York, Oxford: Oxford University Press.

Anctil, Michel. 2015. *Dawn of the Neuron: The Early Struggles to Trace the Origin of Nervous Systems*. Montreal: McGill-Queen's University Press.

– 2018. *Luminous Creatures: The History and Science of Light Production in Living Organisms*. Montreal: McGill-Queen's University Press.

Anonymous. 1884. "Bergmann, Karl Georg Lucas Christian." Page 411 in *Biographisches Lexikon der Hervorragenden Aerzte aller Zeiten und Völker*, Volume 1, edited by A. Wernich and August Hirsch. Vienna and Leipzig: Urban & Schwarzenberg.

Appel, Toby A. 1987. *The Cuvier-Geoffroy Debate: French Biology in the Decades before Darwin*. New York, Oxford: Oxford University Press.

Arago, François. 1833. "Eloge historique d'Alexandre Volta, lu à la séance publique du 26 juillet 1831." *Mémoires de l'Académie Royale des Sciences* 12: lvij–civ.

Arp, Alissa J., and James J. Childress. 1983. "Sulfide binding by the blood of the hydrothermal vent tube worm *Riftia pachyptila*." *Science* 219: 295–7.

Arp, Alissa J., James J. Childress, and Russell D. Vetter. 1987. "The sulphide-binding protein in the blood of the vestimentiferan tube-worm, *Riftia pachyptila*, is the extracellular haemoglobin." *Journal of Experimental Biology* 128: 139–58.

Arvanitaki, Angélique, and Henri Cardot. 1941. "Les caractéristiques de l'activité rythmique ganglionnaire 'spontanée' chez l'Aplysie." *Comptes rendus des séances de la Société de Biologie* 135: 1207–11.

Atema, Jelle. 1995. "Chemical signals in the marine environment: Dispersal, detection, and temporal signal analysis." *Proceedings of the National Academy of Sciences of USA* 92: 62–6.

Atwood, Margaret. 2018. "Child Harold." *Journal of Neurogenetics* 32: 131–3.

Atz, James W. 1986. "*Fundulus heteroclitus* in the laboratory: a history." *American Zoologist* 26: 111–20.

Audouin, JeanVictoire, and Henri Milne-Edwards. 1827. *Recherches anatomiques et physiologiques sur la circulation dans les crustacés*. Paris: Imprimerie de C.Thuau.

Autrum, Hansjochem. 1982. "Karl von Frisch: November 20, 1886–June 12, 1983." *Journal of Comparative Physiology* 147: 417–22.

Babkin, Boris P. 1914. *Die äussere Sekretion der Verdauungsdrüsen*. Berlin: Julius Springer.

– 1929. "Studies on the pancreatic secretion in skates." *Biological Bulletin* 57: 272–91.

– 1942. "How I came to Dalhousie." *Journal of the Mount Sinai Hospital* 9: 168–75.

– 1949. *Pavlov: A Biography*. Chicago: University of Chicago Press.

Babkin, Boris P., and D.J. Bowie. 1928. "The digestive system and its function in *Fundulus heteroclitus*." *Biological Bulletin* 54: 254–77.

Babkin, Boris P., D.J. Bowie, and J.V.V. Nicholls. 1933. "Structure and reactions to stimuli of arteries (and conus) in the elasmobranch genus *Raja*." *Contributions to Canadian Biology and Fisheries* 8: 207–25.

Bacq, Zénon M. 1934. "La pharmacologie du système nerveux autonome, et particulièrement du sympathique, d'après la théorie neurohumorale." *Annales de physiologie* 10: 467–553.

– 1935. "La transmission chimique des influx dans le Système nervcux autonome."

Ergebnisse der Physiologie, biologischen Chemie und experimentellen Pharmakologie 37: 82–285.

– 1947. "L'acétylcholine et l'adrénaline chez les invertébrés." *Biological Reviews* 22: 73–91.

– 1975. *Chemical Transmission of Nerve Impuses.: A Historical Sketch*. Oxford, UK: Pergamon Press.

– 1983. "Notice sur Henri Fredericq." *Annuaire de l'Académie Royale de Belgique*: 23–55.

Bacq, Zénon M., and Jean Brachet. 1981. "Notice sur Marcel Florkin." *Annuaire de l'Académie Royale de Belgique*: 41–98.

Baedke, Jan. 2018. "O Organism, where art thou? Old and new challenges for organism-centered biology." *Journal of the History of Biology* 52: 293–324.

Baglioni, Silvestro. 1913. "Die Grundlagen der vergleichenden Physiologie des Nervensystems." Pages 1–450 in *Handbuch der vergleichende Physiologie*, Vol. 4, edited by Hans Winterstein. Jena: Gustav Fischer.

Bailey, David M., Hans-Joachim Wagner, Alan J. Jamieson, Murray F. Ross, and Imants G. Priede. 2007. "A taste of the deep-sea: the roles of gustatory and tactile searching behaviour in the grenadier fish *Coryphaenoides armatus*." *Deep-Sea Research* Part I, 54: 99–108.

Baldwin, Ernest. 1933. "Phosphagen." *Biological Reviews* 8: 74–105.

– 1937. *An Introduction to Comparative Biochemistry*. Cambridge, UK: Cambridge University Press.

– 1947. *Dynamic Aspects of Biochemistry*. Cambridge, UK: Cambridge University Press.

Bange, Christian. 2011. "La physiologie appliquée dans les stations maritimes françaises de biologie entre 1880 et 1930 et les recherches de Raphaël Dubois à Tamaris." *Bulletin mensuel de la Société linnéenne de Lyon* 80: 13–29.

Barbara, Jean-Gaël. 2007. "La neurophysiologie à la française: Alfred Fessard et le renouveau d'une discipline." *Revue pour l'histoire du CNRS* 19: 1–7.

Barham, Eric G. 1970. "Deep-sea fishes: lethargy and vertical orientation." Pages 100–16 in *Proceedings of an International Symposium on Biological Sound Scattering in the Ocean*, edited by G. Brooke Farquhar. Washington, DC: Maury Center for Ocean Science.

Barrington, Ernest J.W. 1975. "Comparative physiology and the challenge of design." *Journal of Experimental Zoology* 194: 271–86.

Barsanti, Giulio. 1994. "Lamarck and the birth of biology, 1740–1810." Pages 47–74 in *Romanticism in Science: Science in Europe*, edited by S. Poggi and M. Bossi. Dordrecht: Kluwer.

Bartholomew, George A. 1942. "The fishing activities of double-crested cormorants on San Francisco Bay." *The Condor* 44: 13–21.

– 1958. "The role of physiology in the distribution of terrestrial vertebrates." Pages 81–95 in *Zoogeography*, edited by Carl L. Hubbs. Washington, DC: American Association for the Advancement of Science.

– 1964. "The roles of physiology and behaviour in the maintenance of homeostasis in the desert environment." *Symposia of the Society for Experimental Biology* 18: 7–29.

Bartholomew, George A., and Herbert H. Caswell. 1951. "Locomotion in kangaroo rats and its adaptive significance." *Journal of Mammalogy* 32: 155–69.

Baskett, Thomas F. 2004. "Robert Hooke and the origins of artificial respiration." *Resuscitation* 60: 125–7.

Battle, Helen I. 1935. "Digestion and digestive enzymes in the herring (*Clupea harengus* L.)." *Journal of the Biological Board of Canada* 1: 145–57.

Bazemore, Alva, Kenneth A.C. Elliott, and Ernst Florey. 1956. "Factor I and γ–aminobutyric acid." *Nature* 178: 1052–3.

Becchi, Antonio. 2009. "The body of the architect. Flesh, bones and forces between mechanical and architectural theories." *Proceedings of the Third International Congress on Construction History*: 151–8.

Beer, Theodor. 1894. "Die Accommodation des Fischauges." *Pflügers Archiv für Physiologie* 58: 523–650.

– 1898. "Die Akkommodation des Auges in der Thierrheihe." *Wiener klinischen Wochenschrift* 42: 942–53.

– 1899. "Vergleichend-physiologische Studien zur Statocystenfunction. I. Ueber den angeblicben Gehörsinn und das angebliche Gehörorgan der Crustaceen." Pflügers Archiv für Physiologie 73: 1–41.

Beer, Theodor, A. Bethe, and J. von Uexküll. 1899. "Vorschläge zu einer objektivierenden Nomenklatur in der Physiologie des Nervensystems." *Biologisches Zentralblatt* 19: 517–21.

Benyus, Janine M. 1997. *Biomimicry: Innovation Inspired by Nature*. New York: Morrow.

Bergman, ke, Jerrold J. Heindel, et al., eds. 2013. *State of the Science of Endocrine Disrupting Chemicals– 2012*. Geneva, Switzerland: United Nations Environment Programme and the World Health Organization.

Bergmann, Karl. 1847. "Über die Verhältnisse der Wärmeökonomie der Thiere zu ihrer Grösse." *Göttinger Studien* 3: 595–708.

Bernard, Claude. 1865. *Introduction à l'étude de la médecine expérimentale.* Paris: J.B. Baillière et Fils.

— 1878. *La Science expérimentale.* Paris: J.B. Baillière et Fils.

Bernardi, Walter. 2001. "La controverse sur l'électricité animale dans l'Italie du XVIIIe siècle: Galvani, Volta et ... d'autres." *Revue d'histoire des sciences* 54: 53–70.

Bert, Paul. 1867. "Sur la physiologie de la seiche (*Sepia officinalis*, L.)." *Mémoires de la Société des sciences physiques et naturelles de Bordeaux* 5(2): 1–4.

— 1870. *Leçons sur la physiologie comparée de la respiration.* Paris: J.B. Baillière et Fils.

Berthelot, Marcelin. 1891. *Notice historique sur Henri Milne-Edwards.* Paris: Typographie de Firmin-Didot et Cie.

Besson, Jacques. 1569. *L'art et science de trouver les eaux et fontaines cachées sous terre, autrement que par les moyens vulgaires des Agriculteurs et Architectes.* Orléans: Pierre Trepperel Libraire.

Bibliowicz, Jonathan, Alexandre Alié, et al. 2013. "Differences in chemosensory response between eyed and eyeless *Astyanax mexicanus* of the Rio Subterráneo cave." *EvoDevo* 4(1): 25.

Bichat, Xavier. 1799. *Traité des membranes en général et de diverses membranes en particulier.* Paris: Richard, Caillé et Ravier.

— 1800. *Recherches physiologiques sur la vie et la mort.* Paris: Brosson, Gabon et Cie.

Black, Edgar C. 1951. "The respiration of fishes." *University of Toronto Studies, Biological Series* 59: 91–111.

Black, Virginia S. 1951. "Osmotic regulations in teleost fishes." *University of Toronto Studies, Biological Series* 59: 53–89.

Bohn, Georges. 1901. "Des mécanismes respiratoires chez les crustacés décapodes." *Bulletin scientifique de la France et de la Belgique* 36: 1–374.

Bolek, Siegfried, Matthias Wittlinger, and Harald Wolf. 2012. "What counts for ants? How return behaviour and food search of *Cataglyphis* ants are modified by variations in food quantity and experience." *Journal of Experimental Biology* 215: 3218–22.

Bone, Quentin, and Nigel R. Merrett. 1998. "Norman Bertram Marshall. 5 February 1915–13 February 1996." *Biographical Memoirs of Fellows of the Royal Society* 44: 281–96.

Boorstin, Daniel J. 1983. *The Discoverers: A History of Man's Search to Know His World and Himself*. New York: Random House.

Borelli, Giovanni Alfonso. 1680. *De motu animalium*. Rome: A. Bernabo. (I consulted Paul Maquet's 1989 English translation, *On the Movement of Animals*, Berlin-Heidelberg: Springer-Verlag.)

Bornancin, Michel. 1980. "Contribution of Jean Maetz to the study of branchial ionic exchange." *American Journal of Physiology* 238: R143–4.

Bottazzi, Filippo. 1897. "La pression osmotique du sang des animaux marins." *Archives italiennes de Biologie* 28: 61–72.

– 1908. "Osmotischer Drück und elektrische Leitfähigkeit der einzelligen, pflanzlichen und tierischen Organismen." *Ergebnisse der Physiologie* 7: 161–402.

Bourdon, Isidore. 1828. *Principes de physiologie médicale*. Paris: J.B. Baillière & Gabon.

– 1830. *Principes de physiologie comparée*. Paris: Gabon & J.B. Baillière.

Boury, Dominique. 2008. "Irritability and sensibility: key concepts in assessing the medical doctrines of Haller and Bordeu." *Science in Context* 21: 521–35.

Bower, Tom. 1992. *Maxwell: The Outsider*. New York: Viking Press.

Boycott, Brian B. 1998. "John Zachary Young 1907–1997." *Biographical Memoirs of Fellows of the Royal Society of London* 44: 485–509.

Brauer, August. 1908. *Wissenschaftliche Ergebnisse der Deutschen Tiefsee-Expedition auf dem Dampfer "Valdivia" 1898–1899. Fünfzehnter Band: Die Tiefsee-Fische. 2: Anatomischer Teil*. Jena: Verlag von Gustav Fischer.

Braun, Marta. 1992. *Picturing Time: The Work of Etienne-Jules Marey (1830–1904)*. Chicago: University of Chicago Press.

Bretag, Allan H. 2017. "The glass micropipette electrode: A history of its inventors and users to 1950." *Journal of General Physiology* 149: 417–30.

Brett, J. Roland. 1956. "Some principles in the thermal requirements of fishes." *Quarterly Review of Biology* 31: 75–87.

– 1964. "The respiratory metabolism and swimming performance of young sockeye salmon." *Journal of the Fisheries Research Board of Canada* 21: 1183–226.

– 1979. Quoted in *This Week's Citation Classic* 17, 23 April 1979.

Brillat-Savarin, Jean Anthelme. 1826. *Physiologie du goût, ou méditations de gastronomie transcendante*. Paris: A. Sautelet.

Brock, Arthur John (Translator). 1916. *Galen on the Natural Faculties*. Cambridge, Mass.: Harvard University Press.

Brown, Margaret E. 1946. "The growth of brown trout (Salmo trutta Linn.). I.

Factors influencing the growth of trout fry." *Journal of Experimental Biology* 22: 118–29.

– 1957. *The Physiology of Fishes.* Volumes I and II. New York: Academic Press.

Bückmann, Detlef. 1985. "Wolfgang von Buddenbrock und die Begründung der vergleichenden Physiologie." *Medizinhistorisches Journal* 20: 120–34.

Buddenbrock, Wolfgang von. 1928. *Grundriss der vergleichenden Physiologie.* Berlin: Gebrüder Borntraeger.

Bullock, Theodor H. 1940. "The functional organization of the nervous system of Enteropneusta." *Biological Bulletin* 79: 91–113.

– 1944. "The giant nerve fibre system in balanoglossids." *Journal of Comparative Neurology* 80: 355–67.

– 1946. "A preparation for the physiological study of the unit synapse." *Nature* 158: 555–6.

– 1977. "Some perspectives on comparative neurophysiology." Pages 533–8 in *Identified Neurons and Behavior in Athropods*, edited by Graham Hoyle. New York and London: Plenum Press.

– 1986. "'Simple' model systems need comparative studies: differences are as important as commonalities" *Trends in Neuroscience* 9: 470–2.

– 1993. *How Do Brains Work? Papers of a Comparative Neurophysiologist.* New York: Springer Science+Business Media.

– 1996. "Theodor H. Bullock." Pages 112–56 in *The History of Neuroscience in Autobiography*, Volume 1, edited by Larry R. Squire. Washington, DC: Society for Neuroscience.

Bullock, Theodor H., S. Hagiwara, K. Kusano, and K. Negishi. 1961. "Evidence for a category of electroreceptors in the lateral line of gymnotid fishes." *Science* 134: 1426–7.

Burggren, Warren W., and Stephen Warburton. 2005. "Comparative developmental physiology: an interdisciplinary convergence." *Annual Review of Physiology* 67: 203–23.

Burgh Daly, Ivan, S.A. Komarov, and E.G. Young. 1952. "Boris Petrovitch Babkin 1877–1950." *Biographical Memoirs of the Royal Society of London* 8: 13–23.

Burne, Richard H., and John R. Norman. 1943. "Charles Tate Regan 1878–1943." *Obituary Notices of Fellows of the Royal Society* 4: 411–26.

Burrows, Malcolm. 2004. "Memories of Bob Boutilier in Cambridge." *Journal of Experimental Biology* 207: 1052.

Butenandt, Adolf. 1959. "Wirkstoffe des Insektenreiches." *Naturwissenschaften* 46: 461–71.

Bynum, William F. 1994. *Science and the Practice of Medicine in the Nineteenth Century*. Cambridge, UK: Cambridge University Press.

Campbell, Gordon L. 2014. "Introduction," Pages 1–4 in *The Oxford Handbook of Animals in Classic Thought and Life*, edited by Gordon L. Campbell. Oxford, UK: Oxford University Press.

Candèze, Ernest. 1874. "Les moyens d'attaque et de défense chez les insectes." *Bulletins de l'Académie royale des sciences, des lettres et des beaux-arts de Belgique*, series 2, 38: 787–816.

Canguilhem, Georges. 2008. *Knowledge of Life*. Edited by Paola Marrati and Todd Meyers. New York: Fordham University Press.

Cannon, Walter B. 1915. *Bodily Changes in Pain, Hunger, Fear and Rage*. New York and London: D. Appleton and Company.

– 1922. "Henry Pickering Bowditch." *Biographical Memoirs of the National Academy of Sciences* 17: 183–96.

– 1931. "Recent studies on chemical mediations of nerve impulses." *Endocrinology* 15: 473–80.

– 1932. *The Wisdom of the Body*. New York: W.W. Norton & Company.

– 1943. "Lawrence Joseph Henderson 1878–1942)." *Biographical Memoirs of the National Academy of Sciences* 23: 31–58.

– 1945. *The Way of an Investigator*. New York: W.W. Norton & Company.

Cannon, Walter B., and Zénon M. Bacq. 1931. "Studies on the conditions of activity in endocrine organs. XXVI. A hormone produced by sympathetic action on smooth muscle." *American Journal of Physiology* 96: 392–412.

Carr, Archie F. 1967. *So Excellent a Fishe: A Natural History of Sea Turtles*. Garden City, NY: Natural History Press.

Carus, Paul. 1894. "The seat of consciousness." *Journal of Comparative Neurology* 4: 176–92.

Castellucci, Vincent, and Eric R. Kandel. 1976. "Presynaptic facilitation as a mechanism for behavioral sensitization in Aplysia." *Science* 194: 1176–8.

Castellucci, Vincent, H. Pinsker, H. Kupfermann, and Eric R. Kandel. 1970. "Neuronal mechanisms of habituation and dishabituation of the gill-withdrawal reflex in *Aplysia*." *Science* 167: 1745–8.

Castillo, José del, Graham Hoyle, and Xenia Machne. 1953. "Neuromuscular transmission in a locust." *Journal of Physiology* 121: 539–47.

Cathcart, Edward P. 1922. "Prof. Max Verworn." *Nature* 109: 213.

Cavanaugh, Colleen M., Stephen L. Gardiner, Meredith L. Jones, Holger W. Jannasch, and John B. Waterbury. 1981. "Prokaryotic cells in the hydrothermal vent tube worm *Riftia pachyptila* Jones: possible chemoautotrophic symbionts." *Science* 213: 340–2.

Charlton, Milton P. 2018. "Fifty years my mentor: Harold Atwood." *Journal of Neurogenetics* 32: 134–41.

Chavot, Philippe. 1994. *Histoire de l'éthologie. Recherches sur le développement des sciences du comportement en Allemagne, Grande-Bretagne et France, de 1930 à nos jours.* Thèse de doctorat, Université Louis Pasteur (Strasbourg I).

Chester-Jones, Ian, P.M. Ingleton, and J.G. Phillips. 1987. *Fundamentals of Comparative Vertebrate Endocrinology.* New York: Springer Science+Business Media.

Chevreuil, Michel-Eugène. 1866. *Histoire des connaissances chimiques.* Paris: L. Guérin.

Childress, James J. 1968. "Oxygen minimum layer: vertical distribution and respiration of the mysid *Gnathophausia ingens.*" *Science* 160: 1242–3.

– 1971. "Respiratory rate and depth of occurrence of midwater animals." *Limnology and Oceanography* 16: 104–6.

Childress, James J., and Charles R. Fisher. 1992. "The biology of hydrothermal vent animals: physiology, biochemistry, and autotrophic symbioses." *Oceanography and Marine Biology Annual Review* 30: 337–441.

Childress, James J., and Mary Nygaard. 1973. "The chemical composition of midwater fishes as a function of depth of occurrence off southern California." *Deep-Sea Research* 20: 1093–109.

– 1974. "Chemical composition and buoyancy of midwater crustaceans as function of depth of occurrence off Southern California." *Marine Biology* 27: 225–38.

Clarac, François, and Edouard Pearlstein. 2007. "Invertebrate preparations and their contribution to neurobiology in the second half of the 20th century." *Brain Research Reviews* 54: 113–61.

Cloudsley-Thompson, John L. 1991. *Ecophysiology of Desert Arthropods and Reptiles.* Berlin, Heidelberg: Springer-Verlag.

Collip, James B. 1920. "The alkali reserve of marine fish and invertebrates." *Journal of Biological Chemistry* 44: 329–44.

– 1925. "The effect of insulin on the oxygen consumption of certain marine fish and invertebrates." *American Journal of Physiology* 72: 181.

Cook, John S. 1987. "Sectionalization." Pages 435–61 in *History of the American*

Physiological Society. The First Century, 1887–1987, edited by John R. Brobeck, Orr E. Reynolds, and Toby A. Appel. New York: Springer.

Cooper, J.C., A.T. Scholz, R.M. Horral, A.D. Hasler, and D.M. Madison. 1976. "Experimental confirmation of the olfactory hypothesis with homing, artifically imprinted coho salmon (*Oncorhynchus kisutch*)." *Journal of the Fisheries Research Board of Canada* 33: 703–10.

Corsi, Pietro. 1987. "Julien Joseph Virey, le premier critique de Lamarck." Pages 181–92 in *Histoire du concept d'espèce dans les sciences de la vie*, edited by Scott Atran. Paris: Editions de la Fondation Singer-Polignac.

Crane, Kathleen. 2003. *Sea Legs: Tales of a Woman Oceanographer*. Boulder, Colorado: Westview Press.

Crane, Kathleen, and William R. Normak. 1977. "Hydrothermal activity and crestal structure of the East Pacific Rise at 21°N." *Journal of Geophysical Research* 82: 5336–48.

Cranefield, Paul F., editor. 1982. *Two Great Scientists of the Nineteenth Century: Correspondence of Emil du Bois-Reymond and Carl Ludwig*. Baltimore: Johns Hopkins University Press.

Crosland, Maurice. 1992. *Science under Control: The French Academy of Sciences 1795–1914*. Cambridge, UK: Cambridge University Press.

Culver, David C., and Tanja Pipan. 2009. *The Biology of Caves and Other Subterranean Habitats*. Oxford, UK: Oxford University Press.

Cunningham, Andrew. 2002. "The pen and the sword: recovering the disciplinary identity of physiology and anatomy before 1800. I: Old physiology – the pen." *Studies in History and Philosophy of Biological and Biomedical Sciences* 33: 631–65.

Curtis, D.R., J.W. Phyllis, and J.C. Watkins. 1960. "The chemical excitation of spinal neurones by certain acidic amino acids." *Journal of Physiology* 150: 656–82.

Dale, Henry H. 1961. "Thomas Renton Elliott, 1877–1961." *Biographical Memoirs of Fellows of the Royal Society* 7: 52–74

Danovaro, Roberto et al. 2010. "The first metazoa living in permanently anoxic conditions." *BMC Biology* 8:30.

Darwin, Horace. 1919. "Keith Lucas 1879–1916." *Proceedings of the Royal Society of London. Series B*, 90(634): xxxi–xxxviii.

Dason, Jeffrey S., Maria P. Sokolowski, and Chun-Fang Wu. 2018. "A reductionist approach to understanding the nervous system: the Harold Atwood legacy." *Journal of Neurogenetics* 32: 127–30.

Dawson, Percy M. 1906. "A biography of François Magendie." *Medical Library and Historical Journal* 4: 45–56.

Dawson, William R. 2007. "Lawrence Irving: an appreciation." *Physiological and Biochemical Zoology* 80: 9–24.

– 2011. *George A. Bartholomew, 1919–2006. A Biographical Memoir.* Washington, DC: National Academy of Sciences.

De Leo, Angela. 2008. "Enrico Sereni: research on the nervous system of cephalopods." *Journal of the History of the Neurosciences* 17:56–71.

Debaz, Josquin. 2005. *Les stations françaises de biologie marine et leurs périodiques entre 1872 et 1914.* Doctoral Thesis, Ecole des Hautes Etudes en Sciences Sociales, Paris.

Dechambre, Amédée. 1884. "Dugès (Antoine-Louis)." *Dictionnaire encyclopédique des sciences médicales* 64: 642–4.

Delaunay, Henri. 1931. "L'excrétion azotée des invertébrés." *Biological Reviews* 6: 265–301.

Délye, Gérard. 1967. "Observations sur la fourmi saharienne *Cataglyphis bombycina* Rog." *Insectes Sociaux* 4: 77–83.

Denton, Eric J., and Norman B. Marshall. 1958. "The buoyancy of bathypelagic fishes without a gas-filled swimbladder." *Journal of the Marine Biological Association of the United Kingdom* 37: 753–67.

Denver, Robert J., Penny M. Hopkins, et al. 2009. "Comparative endocrinology in the 21st century." *Integrative and Comparative Biology* 49: 339–48.

Descarries, Laurent, and Denis Umbriaco. 1995. "Ultrastructural basis of monoamine and acetylcholine function in CNS." *Seminars in the Neurosciences* 7: 309–18.

Descartes, René. 1677. *L'homme et la formation du foetus.* Paris: Théodore Girard.

DeVries, Arthur L., and Donald E. Wohlschlag. 1969. "Freezing resistance in some Antarctic fishes." *Science* 163: 1073–5.

Diedrich, Maria L. 2010. *Cornelia James Cannon and the Future American Race.* Amherst and Boston: University of Massschusetts.

Dijkgraaf, Sven. 1960. "Spallanzani's unpublished experiments on the sensory basis of object perception in bats." *Isis* 51: 9–20.

Douglas, Ron H., and Julian C. Partridge. 1997. "On the visual pigments of deep-sea fish." *Journal of Fish Biology* 50: 68–85.

Du Bois-Reymond, Emil. 1848. *Untersuchungen über thierische Elektricität*, Erster Band. Berlin: Verlag von G. Reimer.

– 1852. *On Animal Electricity: Being an Abstract of the Discoveries of Emil du Bois-Reymond*. Edited by H. Bence Jones. London: John Churchill.

– 1874. "Limits of our knowledge of nature." *Popular Science Monthly* 5: 17–32.

Dubuisson, Marcel. 1931. "Contribution à l'étude de la physiologie du muscle cardiaque des Invertébrés. 7. L'automatisme et le rôle du plexus nerveux cardiaque de *Limulus polyphemus*. *Archives internationals de Physiologie* 33: 257–72.

Ducrotay de Blainville, Henri. 1833. *Cours de physiologie générale et comparée*. Paris: Germer Baillière.

Dugès, Antoine. 1938. *Traité de physiologie comparée de l'homme et des animaux*. Tomes 1–3. Montpellier: Louis Castel; Paris: J.-B. Baillière.

Dupont, Jean-Claude. 1999. *Histoire de la neurotransmission*. Paris: Presses universitaires de France.

Dzendolet, Ernest. 1967. "Behaviorism and sensation in the paper by Beer, Bethe, and von Uexküll (1899)." *Journal of the History of the Behavioral Sciences* 3: 256–61.

Egerton, Frank N. 2016. "History of ecological sciences, Part 56: Ethology until 1973." *Bulletin of the Ecological Society of America* 97: 31–88.

Eigenmann, Carl H. 1909. *Cave Vertebrates of America: A Study in Degenerative Evolution*. Washington, DC: Carnegie Institution of Washington.

Eisner, Thomas, and Jerrold Meinwald. 1966. "Defensive Secretions of Arthropods." *Science* 153: 1341–50.

Elliott, Thomas R. 1904. "On the action of adrenalin." *Journal of Physiology* 31(supplement): xx–xxi.

Erlingsson, Steindór J. 2009. "The Plymouth Laboratory and the institutionalization of experimental zoology in Britain in the 1920s." *Journal of the History of Biology* 42: 151–83.

– 2013. "Institutions and innovation: experimental zoology and the creation of the British Journal of Experimental Biology and the Society for Experimental Biology." *British Journal of the History of Science* 46: 73–95.

– 2016. "'Enfant terrible': Lancelot Hogben's life and work in the 1920s." *Journal of the History of Biology* 49: 495–526.

Euler, Ulf S. von. 1946. "A specific sympathomimetic ergone in adrenergic nerve fibres (sympathin) and its relations to adrenaline and nor-adrenaline." *Acta Physiologica Scandinavia* 12: 73–97.

Euler, Ulf S. von, and H. Heller, eds. 1963. *Comparative Endocrinology. Volume 1:*

Glandular Hormones. Volume 2: Invertebrate and Tissue Hormones. New York and London: Academic Press.

Evans, David O., and William H. Neill. 1990. "Introduction to the Proceedings of the Symposium 'From environment to fish to fisheries: a tribute to F.E.J. Fry.'" *Transactions of the American Fisheries Society* 119: 567–70.

Evans, David O., A.M. McCombie, and J.M. Casselman. 1990. "In memoriam: Frederick Ernest Joseph Fry, 1909–1989." *Transactions of the American Fisheries Society* 119: 571–3.

Exner, Sigmund. 1891. *Die Physiologie der Facettirten Augen von Krebsen und Insecten.* Leipzig und Wien: Franz Deuticke.

Fabre, Jean-Henri. 1913. *Insect Life: Souvenirs of a Naturalist.* London: Macmillan and Co.

Fenchel, Tom, and Bland J. Finlay. 1995. *Ecology and Evolution in Anoxic Worlds.* Oxford and New York: Oxford University Press.

Figard, Léon. 1903. *Un médecin philosophe au XVIe siècle. Etude sur la psychologie de Jean Fernel.* Paris: Félix Alcan.

Figlio, Karl. 1977. "The historiography of scientific medicine: an invitation to the human sciences." *Comparative Studies in Society and History* 19: 262–86.

Finger, Stanley. 2000. *Minds Behind the Brain: A History of the Pioneers and Their Disoveries.* New York: Oxford University Press.

Finkelstein, Gabriel. 2013. *Emil du Bois-Reymond: Neuroscience, Self, and Society in Nineteenth-Century Germany.* Cambridge, Mass.: The MIT Press.

Fischer, Jean Louis. 1980. "L'aspect social et politique des relations épistolaires entre quelques savants français et la Station zoologique de Naples de 1878 à 1912." *Revue d'histoire des sciences* 33: 225–51.

Fisher, Charles R., James J. Childress, and Elizabeth Minnich. 1989. "Autotrophic carbon fixation by the chemoautotrophic symbionts of *Riftia pachyptila.*" *Biological Bulletin* 177: 372–85.

Florey, Ernst. 1984. "Synaptic and nonsynaptic transmission: a historical perspective." *Neurochemical Research* 9: 413–26.

– 1990. "Crustacean neurobiology: history and perspective." Pages 4–32 in *Frontiers in Crustacean Neurobiology,* edited by K. Wiese, W.-D. Krenz, J. Tautz, H. Reichert, and B. Mulloney. Basel: Springer Basel AG.

Florkin, Marcel. 1944. *L'évolution biochimique.* Liège: Editions Desoer.

– 1952. "Comparative biochemistry." *Annual Review of Biochemistry* 21: 459–72.

– 1973. "The call of comparative biochemistry." *Comparative Biochemistry and Physiology* 44B: 1–10.

– 1979. *L'Ecole liégeoise de physiologie et son maître Léon Frédéricq (1851–1935), pionnier de la zoologie chimique.* Liège: Vaillant-Carmanne.

Florkin, Marcel, and Ernest Schoffeniels. 1969. *Molecular Approaches to Ecology.* New York and London: Academic Press.

Flourens, Marie-Jean-Pierre. 1824. *Recherches expérimentales sur les propriétés et les fonctions du système nerveux dans les animaux vertébrés.* Paris: Crevot.

– 1840. "Eloge historique de F. Cuvier." *Mémoires de l'Académie des sciences de l'Institut de France* 18: ij–xviij.

– 1854. "Eloge historique de Marie-Henri Ducrotay de Blainville" *Mémoires de l'Académie des sciences de l'Institut de France* 25: i–lx.

– 1856. *Cours de physiologie comparée. De l'ontologie ou étude des êtres.* Paris: J.B. Baillière.

Fontaine, Maurice. 1936. "Sur la maturation complète des organes génitaux de l'anguille mâle et l'émission spontanée de ses produits sexuels." *Comptes rendus hebdomadaires des séances de l'Académie des Sciences* 202: 1312–4.

– 1944. "Quelques données récentes sur le mécanisme physiologique des migrations de l'anguille européenne." *Bulletin français de pisciculture* 134: 5–19.

– 1954. "Du déterminisme physiologique des migrations." *Biological Reviews* 29: 390–418.

Fontaine, Maurice, Aubrey Gorbman, J. Leloup, and M. Olivereau. 1952. "La glande thyroïde de la Lamproie marine (Petromyzon marinus L.) lors de sa montée reproductrice." *Annales d'Endocrinologie* 13: 55–7.

Foster, Michael. 1901. *Lectures on the History of Physiology.* London, UK: C.J. Clay and Sons.

Frazier, Wesley T., Eric R. Kandel, Irving Kupfermann, Rafiq Waziri, and Richard E. Coggeshall. 1967. "Morphological and functional properties of identified neurons in the abdominal ganglion of *Aplysia californica.*" *Journal of Neurophysiology* 30: 1288–351.

Frederick II of Prussia. 1752. *Eloge de La Mettrie.* The Hague, Netherland.

Fredericq, Henri. 1913. "L'excitabilité du vague cardiaque et ses modifications sous l'influence de la caféine." *Archives internationales de Physiologie* 13: 107–25.

– 1930. "Action des nerfs inhibiteurs sur la chronaxie cardiaque, nerveuse et musculaire de *Limulus polyphemus.*" *Archives Internationales de Physiologie* 32: 126–37.

– 1947. "Les nerfs cardio-régulateurs des invertébrés et la théorie des médiateurs chimiques." *Biological Reviews* 22: 297–314.

Fredericq, Léon. 1878. "Recherches sur la physiologie du poulpe commun (*Octopus vulgaris*)." *Archives de zoologie expérimentale et générale* 7: 535–83.

– 1883. "Sur l'autotomie, ou mutilation par voie réflexe comme moyen de défense chez les animaux." *Archives de zoologie expérimentale et générale* 11: 413–26.

– 1891. "Sur la physiologie de la branchie." *Archives de zoologie expérimentale et générale* 19: 117–24.

– 1892. *Manipulations de physiologie*. Paris: J.B. Baillière et Fils.

– 1901. "Sur la concentration moléculaire du sang et des tissus chez les animaux aquatiques." *Bulletin de l'Académie Royale des Sciences de Belgique* (1901): 428–54.

Freeman, James A. 1963. "Review: A textbook of comparative endocrinology, Aubrey Gorbman and Howard A. Bern." *The Anatomical Record* 145: 473–5.

Fridberg, Gunnar, and Howard A. Bern. 1968. "The urophysis and the caudal neurosecretory system of fishes." *Biological Reviews* 43: 175–99.

Frisch, Karl von. 1915. "Die Farbensinn und Formensinn der Biene." *Zoologische Jahrbücher* 35: 2–182.

– 1919. "Über den Geruchsinn der Biene und seine blütenbiologische Bedeutung." *Zoologische Jahrbücher* 37: 1–26.

– 1923. "Ein Zwergwels, der kommt, wenn man ihm pfeift." *Biologisches Zentralblatt* 43: 439–46.

– 1927. *Aus dem Leben der Bienen*. Berlin: Springer-Verlag.

– 1938. "Zur Psychologie des Fisch-Schwarmes." *Naturwissenschaften* 26: 601–6.

– 1962. *Erinnerungen eines Biologen*. Berlin, Heidelberg: Springer-Verlag.

Fry, Frederick E.J. 1947. "Effects of the environment on animal activity." *University of Toronto Studies, Biological Series* 55: 1–62.

– 1958. "Temperature compensation." *Annual Review of Physiology* 20: 207–24.

Fry, Frederick E.J., and J. Sanford Hart. 1948a. "Cruising speed of goldfish in relation to water temperature." *Journal of the Fisheries Research Board of Canada* 7: 169–75.

– 1948b. "The relation of temperature to oxygen consumption in the goldfish." *Biological Bulletin* 94: 66–77.

– Virginia S. Black, and Edgar C. Black. 1947. "Influence of temperature on the asphyxiation of young goldfish (*Carassius auratus* L.) under various tensions of oxygen and carbon dioxide." *Biological Bulletin* 92: 217–24.

Fry, Iris. 1996. "On the biological significance of the properties of matter: L J. Henderson's theory of the fitness of the environment." *Journal of the History of Biology* 29: 155–96.

Fürth, Otto von. 1903. *Vergleichende chemische Physiologie der niederen Tiere*. Jena: Gustav Fischer.

Fye, W. Bruce. 1986. "Carl Ludwig and the Leipzig Physiological Institute: 'a factory of knowledge.'" *Circulation* 74: 920–8.

Galambos, Robert, and Donald R. Griffin. 1942. "Obstacle avoidance by flying bats: the cries of bats." *Journal of Experimental Zoology* 89: 475–90.

Galvani, Luigi. 1791. *De viribus electricitatis in motu musculari commentarius*. Bologna: Instituti Scientarum.

– 1797. *Memorie sull'elettricità animale di Luigi Galvani p. Professore di Notomia nella Università di Bologna al celebre Abate Lazzaro Spallanzani Pubblico professore nella Università di Pavia. Aggiunte alcune elettriche esperienze di Gio. Aldini P. prof. di Fisica*. Bologna: Sassi.

Gans, Carl. 1961. "The feeding mechanism of snakes and its possible evolution." *American Zoologist* 1: 217–27.

Garland, Theodore, Jr, Albert F. Bennett, and Enrico L. Rezende. 2005. "Phylogenetic approaches in comparative physiology." *Journal of Experimental Biology* 208: 3015–35.

Gartner, John V., Jr, Roy E. Crabtree, and Kenneth J. Sulak. 1997. "Feeding at depth." Pages 115–93 in *Fish Physiology, Vol. 16: Deep-Sea Fishes*, edited by David J. Randall and Anthony P. Farrell. San Diego, California: Academic Press.

Gatti, Roberto Cazzolla. 2016. "Self-consciousness: beyond the looking-glass and what dogs found there." *Ethology Ecology and Evolution* 28: 232–40.

Gauthier-Clerc, Michel, Yvon Le Maho, Yannick Clerquin, Samuel Drault, and Yves Handrich. 2000. "Penguin fathers preserve food for their chicks." *Nature* 408: 928–9.

Gay, Peter. 1998. *Freud: A Life for Our Time*. New York: W.W. Norton & Company.

Gayon, Jean. 2005. "De la biologie comme science historique." *Les Temps Modernes* 2005/2 (n° 630–1): 55–67.

Geenen, Vincent. 2015. "L'école liégeoise de physiologie aux XIXème et XXème siècles." *Histoire des Sciences Médicales* 44: 209–18.

Geison, Gerald L. 1978. *Michael Foster and the Cambridge School of Physiology*. Princeton, New Jersey: Princeton University Press.

Geoffroy Saint-Hilaire, Etienne. 1822. "La zoologie a-t-elle dans l'Académie des sciences une représentation suffisante? La physiologie n'y·a-t-elle pas été oubliée?" *Revue encyclopédique* 13: 501–11.

Georges-Berthier, Auguste. 1914. "Le mécanisme cartésien et la physiologie au XVIIe siècle." *Isis* 2: 37–89.

Ghiretti, Francesco. 1985. "Comparative physiology and biochemistry at the Zoological Station of Naples." *Biological Bulletin* 168: 122–6.

Gilles, Raymond, and Ernest Schoffeniels. 1969. "Isosmotic regulation in isolated surviving nerves of *Eriocheir sinensis* Milne Edwards." *Comparative Physiology and Biochemistry* 31: 927–39.

Godfrey-Smith, Peter. 2016. *Other Minds: The Octopus, the Sea, and the Deep Origins of Consciousness*. New York: Farrar, Straus and Giroux.

Gorbman, Aubrey. 1941. "Identity of an iodine-storing tissue in an Ascidian." *Science* 94: 192–3.

– 1993. "A parting word," *General and Comparative Endocrinology* 89: 2–3.

Gorbman, Aubrey, and Howard A. Bern. 1962. *A Textbook of Comparative Endocrinology*. New York: John Wiley and Sons.

Gorbman, Aubrey, and Charles W. Creaser. 1942. "Accumulation of radio-active iodine by the endostyle of larval lampreys and the problem of homology of the thyroid." *Journal of Experimental Zoology* 89: 391–405.

Gould, Stephen J. 1977. *Ontogeny and Phylogeny*. Cambridge, Massachusetts: Harvard University Press.

Graham, Judith, and Ralph W. Gerard. 1946. "Membrane potentials and excitation of impaled single muscle fibers." *Journal of Cellular and Comparative Physiology* 28: 99–117.

Grant, V. 1963. *The origin of adaptations*. New York: Columbia University Press.

Grassle, J. Frederick et al. 1979. "Galápagos '79: initial findings of a deep-sea biological quest." *Oceanus* 22: 4–10.

Greenberg, Michael J., Peter W. Hochachka, and Charlotte P. Mangum. 1975. "Biographical data: Clifford Ladd Prosser." *Journal of Experimental Zoology* 194: 5–12.

Griffin, Donald R. 1953. "Acoustic orientation in the oil bird, *Steatornis*." *Proceedings of the National Academy of Science* 39: 884–93.

– 1958. *Listening in the Dark*. New Haven, Connecticut: Yale University Press.

– 1981. "*The Question of Animal Awareness*. New York: Rockefeller University Press.

– 1984. *Animal Thinking*. Cambridge, Massachusetts: Harvard University Press.

– 1992. *Animal Minds*. Chicago: University of Chicago Press.

– 1998. "Donald R. Griffin." Pages 70–93 in *The History of Neuroscience in Autobiography*, Volume 2, edited by Larry R. Squire. San Diego, California: Academic Press.

Griffin, Donald R., and Robert Galambos. 1941. "The sensory basis of obstacle avoidance by flying bats." *Journal of Experimental Zoology* 86: 481–506.

Gudernatsch, J. Frederick. 1912. "Feeding experiments on tadpoles. I. The influence of specific organs given as food on growth and differentiation." Archiv für Entwicklungsmechanik der Organismen 35: 457–83.

Guerrini, Anita. 2016. "Animal experiments and the antivivisection debates in the 1820s." Pages 71–85 in *Frankenstein's Science: Experimentation and Discovery in Romantic Culture, 1780–1830*, edited by Christa Knellwolf and Jane Goodall. Aldershot, UK: Ashgate Publishing Company.

Hagen, Joel B. 2015. "Camels, cormorants, and kangaroo rats: integration and synthesis in organismal biology after World War II." *Journal of the History of Biology* 48: 169–99.

Hagiwara, Susumu, and Theodor H. Bullock. 1957. "Intracellular potentials in pacemaker and integrative neurons of the lobster cardiac ganglion." *Journal of Cellular and Comparative Physiology* 50: 25–47.

Hagner, Michael. 2003. "Scientific medicine." Pages 49–87 in *From Natural Philosophy to the Sciences*, edited by David Cahan. Chicago and London: University of Chicago Press.

Haldane, John S. 1917. *Organism and Environment as Illustrated by the Physiology of Breathing*. New Haven: Yale University Press.

Haller, Albrecht von. 1751. *Primae lineae physiologiae in usum praelectionum academicarum, aucta et emendata*. Göttingen: Ap., Widow of Abraham Vandenhoeck.

– 1755. *Dissertation sur les parties irritables et sensibles des animaux*. Lausanne: Marc-Michel Bousquet et Compagnie.

– 1762. *Elementa physiologiae corporis humani. Tomus IV: Cerebrum, Nervi, Musculi*. Lausanne: Sumptibus Francisci Grasset et Sociorum.

Hamy, Ernest-Théodore. 1894. *Vie et travaux de M. de Quatrefages*. Paris: F. Alcan.

Hanström, Bertil. 1937. "Inkretorische Organe und Hormonfunktionen bei den Wirbellosen." *Ergebnisse der Biologie* 14: 143–224.

– 1939. *Hormones in Invertebrates*. Oxford, UK: Oxford University Press.

Harden Jones, Frederick R., and Norman B. Marshall. 1953. "The structure and function of the teleostean swimbladder." *Biological Reviews* 28: 16–82.

Harreveld, Anthonie van, and M. Mendelson. 1958. "Glutamate-induced contractions in crustacean muscle." *Journal of Cellular and Comparative Physiology* 54: 85–94.

Hart, J.L. 1958. "Biological Station, St. Andrews, N.B. 1908–1958: Fifty Years of Research in Aquatic Biology." *Journal of the Fisheries Research Board of Canada* 15: 1127–61.

Hart, J. Sanford. 1943. "The cardiac output of four freshwater fish." *Canadian Journal of Research* 21: 77–84.

Harvey, William. 1628. *Exercitatio anatomica de motu cordis et sanguinis in animalibus. Tercentennial Edition.* Springfield, Illinois and Baltimore, Maryland: Charles C. Thomas.

Hasler, Arthur D., and Allan T. Scholz. 1983. *Olfactory Imprinting and Homing in Salmon.* Berlin, Heidelberg: Springer-Verlag.

Hayashi, Takashi. 1954. "Effects of sodium glutamate on the nervous system." *Keio Journal of Medicine* 3: 183–92.

Haydon, Philip G., and Pierre Drapeau. 1995. "From contact to connection: early events during synaptogenesis." *Trends in Neuroscience* 18: 196–201.

Heinrich, Bernd. 1970. "Thoracic temperature stabilization by blood circulation in a free-flying moth." *Science* 168: 580–2.

– 1972. "Energeties of temperature regulation and foraging in a bumblebee, *Bombus terricola* Kirby." *Journal of Comparative Physiology* 77: 49–64.

– 2007. *The Snoring Bird: My Family's Journey through a Century of Biology.* New York: Ecco Press.

Heinrich, Bernd, and Ann E. Kammer. 1973. "Activation of the fibrillar muscles in the bumblebee during warm-up, stabilization of thoracic temperature and flight." *Journal of Experimental Biology* 58: 677–88.

Henderson, Lawrence J. 1913. *The Fitness of the Environment: An Inquiry into the Biological Significance of the Properties of Matter.* New York: Macmillan.

Herrick, C. Judson. 2005. "The central gustatory paths in the brains of bony fishes." *Journal of Comparative Neurology and Psychology* 15: 375–456.

– 1924. *Neurological Foundations of Animal Behavior.* New York: Henry Holt and Company.

Herring, Peter J. 2002. *The Biology of the Deep Ocean.* Oxford, UK: Oxford University Press.

Hesse, Richard. 1924. *Tiergeographie auf ökologischer Grundlage.* Jena: Gustav Fischer.

Hisao, William W.L., Christine Metz, Davinder P Singh, and Jesse Roth. 2008. "The microbes of the intestine: an introduction to their metabolic and signaling capabilities." *Endocrinology and Metabolism Clinics of North America* 37: 857–71.

Hoar, William S. 1937. "The development of the swim bladder of the Atlantic Salmon." *Journal of Morphology* 61: 309–19.

– 1939. "The Thyroid Gland of the Atlantic Salmon." PhD Dissertation, Boston University.

– 1951a. "Hormones in fish." *University of Toronto Studies, Biological Series* 59: 1–51.

– 1951b. "The behaviour of chum, pink and coho salmon in relation to their seaward migration." *Journal of the Fisheries Research Board of Canada* 8: 241–63.

– 1953. "Control and timing of fish migration." *Biological Reviews* 28: 437–52.

– 1966. General and Comparative Physiology. Englewood Cliffs, New Jersey: Prentice-Hall.

– 1982. "Quid usque adhuc? Quo vadimus?" *Canadian Journal of Fisheries and Aquatic Sciences* 39: 155–9.

Hochachka, Peter W. 1962. "Thyroidal effects on pathways for carbohydrate metabolism in a teleost." *General and Comparative Endocrinology* 2: 499–505.

– 1965. "Isoenzymes in metabolic adaptation of a poikilotherm: subunit relationships in lactate dehydrogenases of goldfish." *Archives of Biochemistry and Biophysics* 111: 96–103.

– 1971. "Physiological and biochemical adaptation: an introduction." *American Zoologist* 11:81–2.

Hochachka, Peter W., and George N. Somero. 1973. Strategies of Biochemical Adaptation. Philadelphia: W.B. Saunders.

– 1984. Biochemical Adaptation. Princeton, New Jersey: Princeton University Press.

Hodgkin, Alan L., and Andrew F. Huxley. 1939. "Action potentials recorded from inside a nerve fibre." *Nature* 144: 710–11.

Hogben, Lancelot T. 1924. "The pigmentary effector system IV: A further contribution to the role of pituitary secretion in amphibian colour response." *Journal of Experimental Biology* 1: 249–70.

– 1926. Comparative Physiology. London: Sidgwick & Jackson.

Holmes, Frederic L. 1993. "The old martyr of science: the frog in experimental physiology." *Journal of the History of Biology* 26: 311–28.

Holst, Erich von. 1932. "Untersuchungen über die Funktion des Zentralnervensystems beim Regenwurm." *Zoologische Jahrbücher, Abteilung für allgemeine Zoologie und Physiologie* 51: 547–88.

– 1934. "Studien über Reflexe und Rhythmen beim Goldfisch (Carassius auratus)." *Zeitschrift für vergleichende Physiologie* 20: 582–99.

– 1936. "Über den 'Magnet-Effekt' als koordinierendes Prinzip im Rückenmark." *Pflügers Archiv für die gesamte Physiologie des Menschen und der Tiere* 237: 655–82.

Horvitz, H. Robert, Martin Chalfie, Carol Trent, John E. Sulston, and Peter D. Evans. 1982. "Serotonin and octopamine in the nematode *Caenorhabditis elegans.*" *Science* 216: 1012–14.

Houssay, Bernardo A. 1936. "What we have learned from the toad concerning hypophyseal functions." *New England Journal of Medicine* 214: 913–26.

Hoyle, Graham. 1975. "Identified neurons and the future of neuroethology." *Journal of Experimental Zoology* 194: 51–74.

– 1984. "The scope of neuroethology." *The Behavioral and Brain Sciences* 7: 367–412.

Hoyle, Graham, and Malcolm Burrows. 1973. "Neural mechanisms underlying behavior in the locust *Schistocerca gregaria.* I. Physiology of identified neurons in the metathoracic ganglion." *Journal of Neurobiology* 4: 3–41.

Hoyle, Graham, and Cornelis A.G. Wiersma. 1958. "Excitation at neuromuscular junctions in Crustacea." *Journal of Physiology* 143: 403–25.

Hughes, Graham M., and Cornelis A.G. Wiersma. 1960. "The co-ordination of swimmeret movements in the crayfish, *Procambarus Clarkii* (Girard)." *Journal of Experimental Biology* 37: 657–70.

Humboldt, Alexander von. 1818. *Personal Narrative of Travels to the Equinoctial Regions of the New Continent during the Years 1799–1804.* Vol. III. London: Longman, J. Murray and H. Colburn.

Huntsman, M. Elinor. 1931. "The effect of certain hormone-like substances on the isolated heart of the skate." *Contributions to Canadian Biology and Fisheries* 7: 31–43.

Huxley, Julian. 1912. *The Individual in the Animal Kingdom.* Cambridge, UK: Cambridge University Press.

– 1920. "Metamorphosis of axolotl caused by thyroid-feeding." *Nature* 104: 435.

Huxley, Thomas H. 1874. "On the hypothesis that animals are automata." *Fortnightly Review* 16: 555–80.

Irving, Laurence. 1937. "The respiration of beaver." *Journal of Cellular and Comparative Physiology* 9: 437–51.

– 1972. *Arctic Life of Birds and Mammals Including Man*. New York, Heidelberg, Berlin: Springer-Verlag.

Irving, Laurence, and M.D. Orr. 1935. "The diving habits of the beaver." *Science* 82: 569.

Irving, Laurence, O.M. Solandt, D.Y. Solandt, and K.C. Fisher. 1935a. "Respiratory characteristics of the blood of the seal." *Journal of Cellular and Comparative Physiology* 6: 393–403.

– 1935b. "The respiratory metabolism of the seal and its adjustment to diving." *Journal of Cellular and Comparative Physiology* 7: 137–51.

Israel, Maurice. 2000. "Ladislav Tauc (1926–1999)." *Trends in Neuroscence* 23(2): 47.

James, William. 1879. "Are We Automata?" *Mind* 4(13): 1–22.

Jasper, Herbert H. 1984. "The saga of K.A.C. Elliott and GABA." *Neurochemical Research* 9: 449–60.

Jenkin, Penelope. 1962. *Animal Hormones: A Comparative Survey*. Oxford, UK: Pergamon Press.

Jeuniaux, Charles, and Jacques Balthazart. 1993. "Ernest Schoffeniels (Liège: 11 May 1927–27 July 1992)." *Biochemical Systematics and Ecology* 21: v–viii.

Johansen, Kjell. 1987. "The world as a laboratory: Physiological insights from Nature's experiments." *Advances in Physiological Research*, 1987: 377–96.

Johnstone, Kenneth. 1977. *The Aquatic Explorers: A History of the Fisheries Research Board of Canada*. Toronto and Buffalo: University of Toronto Press.

Johnstone, Rose. 2003. "A sixty-year evolution of biochemistry at McGill University." *Scientia Canadensis* 27: 27–84.

Jordan, Hermann. 1913. *Vergleichende Physiologie wirbelloser Tiere*. Jena: Gustav Fischer.

Kalof, Linda. 2007. *Looking at Animals in Human History*. London, UK: Reaktion Books.

Kandel, Eric R. 2006. *In Search of Memory: The Emergence of a New Science of Mind*. New York and London: W. W. Norton & Company.

Kandel, Eric R., and Ladislav Tauc. 1964. "Mechanism of prolonged heterosynaptic facilitation in a giant ganglion cell of *Aplysia depilans*." *Nature* 202: 145–7.

Karlson, Peter. 1995. "Adolf Butenandt (1903–1995)." *Nature* 373: 660.

Karlson, Peter, and M. Lüscher. 1959. "'Pheromones': new term for a class of biologically active substances." *Nature* 183: 55–6.

Kendall, Edward C., and A.E. Osterberg. 1919. "The chemical identification of thyroxin." *Journal of Biological Chemistry* 40: 265–334.

Kennedy, Donald. 1967. "Small systems of nerve cells." *Scientific American* 216(5): 44–52.

Kerkut, Gerald A. 1970. "Ernest Baldwin, 1909–1969." *Comparative Biochemistry and Physiology* 34: 1–2.

– 1988. "The origins of *Comparative Biochemistry and Physiology*." *Comparative Biochemistry and Physiology* 90A: 1–3.

Kerkut, Gerald A., J.D.C. Lambert, R.J. Gayton, J.E. Locker, and Robert J. Walker. 1975. "Mapping of nerve cells in the suboesophageal ganglia of *Helix aspersa*." *Comparative Biochemistry and Physiology* 50A: 1–25.

Kerr, S.R. 1990. "The Fry paradigm: its significance for contemporary ecology." *Transactions of the American Fisheries Society* 119, 779–85.

Keynes, Richard D. 2005. "J.Z. and the discovery of squid giant nerve fibres." *Journal of Experimental Biology* 208: 179–80.

Kirschner, Leonard B. 1980. "Contribution of Jean Maetz to the physiology of osmotic and ionic regulation in fish." *American Journal of Physiology* 238: R145–6.

Kisling, Vernon N., Jr, editor. 2001. *Zoo and Aquarium History: Ancient Animal Collections in Zoological Gardens*. Boca Raton, Florida: CRC Press.

Klein, Marc. 1954. "Sur l'origine du vocable 'Biologie.'" *Archives d'anatomie, d'histologie et d'embryologie* 37: 105–14.

Knight, Kathryn. 2015. "Celebrating the life and career (to date) of George Somero." *Journal of Experimental Biology* 218: 1799–800.

Knoeff, Rina. 2002. *Herman Boerhaave (1668–1738): Calvinist Chemist and Physician*. Amsterdam: Koninklijke Nederlandse Akademie van Wetenschappen.

Kohler, Robert E. 1982. *From Medical Chemistry to Biochemistry: The Making of a Biomedical Discipline*. Cambridge, UK: Cambridge University Press.

Koller, Gottfried. 1929. "Die innere Sekretion bei wirbellosen Tieren." *Biological Reviews* 4: 269–306.

– 1937. *Hormone bei wirbellosen Tieren*. Leipzig: Akademische Verlagsgesellschaft.

Kooyman, Gerald L., C.M. Drabek, R. Elsner, and W.B. Campbell. 1971. "Diving behavior of the emperor penguin, *Aptenodytes forsteri*." *The Auk* 88: 775–95.

Kooyman, Gerald L., and Paul J. Pontanis. 1998. "The physiological basis of diving to depth: Birds and Mammals." *Annual Review of Physiology* 60: 19–32.

Krebs, Hans A. 1975. "The August Krogh principle: 'For many problems there is an animal on which it can be most conveniently studied.'" *Journal of Experimental Zoology* 194: 221–6.

Krebs, John R., and S. Sjölander. 1992. "Konrad Zacharias Lorenz 7 November

1902–27 February 1989." *Biographical Memoirs of Fellows of the Royal Society* 38: 211–28.

Kristan, William B., Jr, Ronald L. Calabrese, and W. Otto Friesen. 2005. "Neuronal control of leech behavior." *Progress in Neurobiology* 76: 279–327.

Krogh, August. 1929. "The progress of physiology." *American Journal of Physiology* 90: 243–51.

– 1938. "Visual thinking. An autobiographical note." *Organon, Mianowski Institute*, Warsaw, 2: 87–94.

Kruif, Paul de. 1962. *The Sweeping Wind*. New York: Harcourt, Brace & World.

Kuffler, Stephen W., and John G. Nicholls. 1976. *From Neuron to Brain: A Cellular Approach to the Function of the Nervous System*. Sunderland, Massachusetts: Sinauer Associates Inc.

Kühn, Alfred. 1927. " ber den Farbensinn der Bienen." *Zeitschrift für vergleichende Physiologie* 5: 762–800.

Lai, Sandra, Joël Bêti, and Dominique Berteaux. 2017. "Movement tactics of a mobile predator in a meta-ecosystem with fluctuating resources: the arctic fox in the High Arctic." *Oikos* 126: 937–47.

Lamarck, Jean-Baptiste. 1801/2. *Hydrogéologie*. Paris: Agasse & Maillard.

– 1809. *Philosophie zoologique*. Paris: Dentu Libraire.

La Mettrie, Julien Offray de. 1865. *L'homme machine*. Paris: Frédéric Henry.

Landgrebe, Frank W. 1941. "The role of the pituitary and the thyroid in the development of teleosts." *Journal of Experimental Biology* 18: 162–9.

Langley, John N. 1901. "Observations on the physiological action of extracts of the supra-renal bodies." *Journal of Physiology* 27: 237–56.

– 1906. "On nerve endings and on special excitable substances in cells." *Proceedings of the Royal Society* B78: 170–94.

Laporte, Yves. 1998. "Etienne-Jules Marey, founder of the graphic method." *Compte-Rendus de l'Académie des Sciences de Paris, Sciences de la Vie*, 321: 347–54.

Leathes, John B. 1934. "Archibald Byron Macallum, 1858–1934)." Obituary Notices of Fellows of the Royal Society 1: 287–91.

Le Maho, Yvon. 1976. *Thermorégulation et jeûne chez le manchot empereur et le manchot royal*. Thèse de 3e. cycle, Université Claude-Bernard, Lyon.

Le Maho, Yvon, Philippe Delclitte, and Joseph Chatonnet. 1976. "Thermoregulation in fasting emperor penguins under natural conditions." *American Journal of Physiology* 231: 913–22.

Lenoir, Timothy. 1997. *Instituting Science: The Cultural Production of Scientific Disciplines*. Stanford, California: Stanford University Press.

LeRoith, Derek, Joseph Shiloach, Jesse Roth, and Maxine Lesniak. 1980. "Evolutionary origins of vertebrate hormones: substances similar to mammalian insulins are native to unicellular eukaryotes." *Proceedings of the National Academy of Sciences of USA* 77: 6184–8.

Likens, Gene E. 2002. "Arthur Davis Hasler 1908–2001." *Biographical Memoirs of the National Academy of Sciences* 82: 3–14.

Limoges, Camille. 1994. "Milne-Edwards, Darwin, Durkheim and the division of labour: a case study in reciprocal conceptual exchanges between the social and the natural sciences." Pages 317–43 in *The Natural Sciences and the Social Sciences*, edited by I. Bernard Cohen. Dordrecht: Springer Science+Business Media.

Livingstone, Margaret S., Ronald M. Harris-Warrick, and Edward A. Kravitz. 1980. "Serotonin and octopamine produce opposite postures in lobsters." *Science* 208: 76–9.

Lnenicka, Gregory A., and Harold L. Atwood. 1985. "Age-dependent long-term adaptation of crayfish phasic motor axon synapses to altered activity." *Journal of Neuroscience* 5: 459–67.

Locke, Michael. 1996. "Sir Vincent Brian Wigglesworth, C.B.E., 17 April 1899–12 February 1994." *Biographical Memoirs of the Royal Society of London* 42: 540–53.

Locket, N. Adam. 1977. "Adaptations to the deep-sea environment." Pages 68–192 in *The Visual System in Vertebrates*, edited by Frederick Crescitelli. Berlin, Heidelberg: Springer-Verlag.

Loeb, Jacques. 1891. *Untersuchungen zur Physiologichen Morphologie der Thiere*. Wurzburg: Georg Hertz.

– 1898. "Physiology." *Science* 7: 154–6.

Loewi, Otto. 1921. "Uber humorale Übertragbarkeit der Herznervenwirkung. I. Mitteilung." *Pflüger's Archiv für die gesamte Physiologie des Menschen und der Tiere* 189: 239–42.

– 1936. "Quantitative und qualitative Untersuchungen über den Sympathicusstoff." *Pflüger's Archiv für die gesamte Physiologie des Menschen und der Tiere* 237: 504–14.

Loewi, Otto, and Emil Navratil. 1926. "Über humorale Übertragbarkeit der Herznervenwirkung. X. Mitteilung. Über das Schicksal des Vagusstoffs."

Pflüger's Archiv für die gesamte Physiologie des Menschen und der Tiere 214: 678–88.

Lorenz, Konrad. 1935. "Der Kumpan in der Umwelt des Vogels: Der Artgenosse als auslösendes Moment sozialer Verhaltensweisen." *Journal für Ornithologie* 83: 137–213, 289–413.

– 1937. "Über die Bildung des Instinktbegriffes." *Die Naturwissenschaften* 25: 289–300.

– 1962. "Erich von Holst." *Die Naturwissenschaften* 49: 385–6.

– 1981. *The Foundations of Ethology.* New York: Springer Science+Business Media.

Loriaux, D. Lynn. 2016. "Marcello Malpighi (1628–1694). The capillary." Pages 55–7 in *A Biographical History of Endocrinology*, edited by D. Lynn Loriaux. New York: Wiley-Blackwell.

Lower, Richard. 1728. *Tractatus de Corde. Sixth Edition.* Leiden: Johann Herman Verbeek.

Lucas, Keith. 1909a. "The evolution of animal function. Part I." *Science Progress in the Twentieth Century* 3: 472–83.

– 1909b. "The evolution of animal function. Part II." *Science Progress in the Twentieth Century* 4: 321–31.

Ludwig, Carl. 1852. *Lehrbuch der Physiologie des Menschen. Erster Band.* Heidelberg: C.F. Wintersche Verlagshandlung.

Macallum, Archibald B. 1910. "The inorganic composition of the blood in vertebrates and invertebrates, and its origin." *Proceedings of the Royal Society of London* 82: 602–24.

– 1926. "The paleochemistry of the body fluids and tissues." *Physiological Reviews* 6: 316–57.

MacKay, Margaret E. 1922. "The source of insulin: a study of the effect produced on blood sugar by extracts of the pancreas and principal islets of fishes." *Journal of Metabolic Research* 2: 149–72.

– 1929. "Further data concerning the histamine salivary secretion." *American Journal of Physiology* 91: 123–31.

– 1932. "The action of some hormones and hormone-like substances on the circulation in the skate." *Contributions to Canadian Biology and Fisheries* 7: 17–29.

Macleod, John J.R., and W.W. Simpson. 1926. "The immediate post mortem changes in fish muscle." *Contributions to Canadian Biology and Fisheries* 3: 439–56.

Magendie, François. 1825. *Précis élémentaire de physiologie*. Deuxième édition. Paris: Mequignon-Marvis.

Maienschen, Jane. 1987. "Physiology, biology, and the advent of physiological morphology." Pages 177–93 in *Physiology in the American Context*, edited by Gerald L. Geison. New York: Springer.

Mangum, Charlotte, and David Towle. 1977. "Physiological adaptation to unstable environments: inconstancy of the internal milieu in an animal may be a regulatory mechanism." *American Scientist* 65: 67–75.

Mangum, Charlotte, and Peter W. Hochachka. 1998. "New directions in comparative physiology and biochemistry: mechanisms, adaptations, and evolution." *Physiological Zoology* 71: 471–84.

Marey, Etienne-Jules. 1878. *La machine animale: locomotion terrestre et aérienne*. Paris: Germer Baillière et Cie.

– 1885. *La méthode graphique dans les sciences expérimentales*. Paris: G. Masson.

Marshall, Norman B. 1951. "Bathypelagic fishes as sound scatterers in the ocean." *Journal of Marine Research* 10: 1–17.

– 1954. *Aspects of Deep Sea Biology*. London: Hutchinson.

– 1960. "Swimbladder structure of deep-sea fishes in relation to their systematics and biology." *Discovery Report* 31: 1–122.

– 1967. "The olfactory organs of bathypelagic fishes." *Symposia of the Zoological Society of London* 19: 57–70.

– 1979. *Developments in Deep Sea Biology*. Poole, Dorset: Blandford Press.

Marshall, Norman B., and D.W. Bourne. 1964. "A photographic survey of benthic fishes in the Red Sea and Gulf of Aden, with observations on their population density, diversity, and habits." *Bulletin of the Museum of Comparative Zoology* 132: 223–44.

Mast, Samuel O., and C. Ladd Prosser. 1932. "Effect of temperature, salts, and hydrogen ion concentration on rupture of the plasmagel sheet, rate of locomotion, and gel/sol ratio in *Amoeba proteus*." *Journal of Cellular and Comparative Physiology* 1: 333–54.

Mattern, Susan P. 2013. *Prince of Medicine: Galen in the Roman World*. Oxford, New York: Oxford University Press.

Matteucci, Carlo. 1840. *Essai sur les phénomènes électriques des animaux*. Paris: Garilian-Goeuvry et V. Dalmont.

Maynard, Donald M. 1953. "Activity in a crustacean ganglion. I. Cardio-inhibition and acceleration in *Panulirus argus*. *Biological Bulletin* 104: 156–70.

– 1955. "Activity in a Crustacean Ganglion. II. Pattern and Interaction in Burst Formation." *Biological Bulletin* 109: 420–36.

– 1967. "Neural coordination in a simple ganglion." *Science* 158: 531–2.

– 1972. "Simpler networks." *Annals of the New York Academy of Sciences* 193(1): 59–72.

Mazzarello, Paolo. 2009. "Vittorio Erspamer: a student's experience at the Ghislieri College." Pages 37–63 in *In Memory of Vittorio Erspamer*, edited by Lucia Negri, Pietro Melchiorri, Tomas Hökfelt, and Giuseppe Nistico. Rome: Exòrma Edizioni.

McCauley, Robert. 1990. "Frederick Ernest Joseph Fry, M.B.E., M.A., PhD, D.Sc., F.R.S.C. – 1908–1989." *Environmental Biology of Fishes* 27: 241–2.

McEwen, Robert S. 1951. "Charles Gardner Rogers 1875–1950." *Anatomical Record* 109: 552.

McRae, Sandra F. 1987. "A.B. Macallum and physiology at the University of Toronto." Pages 97–114 in *Physiology in the American Context 1850–1940*, edited by Gerald L. Geison. New York: Springer.

Medvei, Victor C. 1982. *A History of Endocrinology*. Lancaster, UK: MTP Press Limited.

Meiri, Shai. 2011. "Bergmann's rule – what's in a name?" *Global Ecology and Biogeography* 20: 203–7.

Meulders, Michel. 2010. *Helmholtz: From Enlightenment to Neuroscience*. Cambridge, Massachusetts: MIT Press.

Mildenberger, Florian. 2006. "The Beer/Bethe/Uexküll paper (1899) and misinterpretations surrounding 'vitalistic behaviorism.'" *History and Philosophy of the Life Sciences* 28: 175–89.

Miller, Frederick R. 1907. "Galvanotropism in the crayfish." *Journal of Physiology* 35: 215–29.

– 1910. "On the rhythmical contractility of the anal musculature of the crayfish and lobster." *Journal of Physiology* 40: 431–44.

Miller, Richard B., A.C. Sinclair, and Peter W. Hochachka. 1959. "Diet, glycogen reserves and resistance to fatigue in hatchery rainbow trout." *Journal of the Fisheries Research Board of Canada* 16: 321–8.

Milne-Edwards, Henri. 1823. "Mémoire sur la structure élémentaire des principaux tissus organiques des animaux." *Archives générales de médecine* 1(3): 165–84.

– 1827a. "Nereis." Pages 529–34 in *Dictionnaire classique d'histoire naturelle, Tome 11*. Paris, Rey et Gravier, Libraires-Editeurs.

– 1827b. "Organisation." Pages 332–44 in *Dictionnaire classique d'histoire naturelle, Tome 12.* Paris, Rey et Gravier, Libraires-Editeurs.

– 1834. *Eléments de zoologie, ou leçons sur l'anatomie, la physiologie, la classification et les moeurs des animaux.* Paris: Chez Crochard.

– 1857. *Leçons sur LA PHYSIOLOGIE et l'anatomie comparée de l'homme et des animaux.* Tome premier. Paris: Librairie de Victor Masson.

Moulins, Maurice, Jean-Pierre Vedel, and Malcolm R. Dando. 1974. "Relations fonctionnelles entre séquences motrices centralement programmées chez les Crustacés décapodes." *Compte-rendus des séances de l'Académie des Sciences* D279: 1895–7.

Müller, Johannes. 1824. "Zur Physiologie des Fötus." *Zeitschrift für die Anthropologie* (1824): 423–83.

– 1826. *Zur vergleichenden Physiologie des Gesichtssinnes.* Leipzig: C. Knobloch,

– 1828. "Ueber ein Eigenthumliches, dem nervus sympathicus analoges Nervensystem der Eingeweide bei den Insecten." *Nova acta physico-medica Academiae Caesareae Leopoldino-Carolinae Naturae Curiosorum* 14, 1–38.

– 1835. "Entdeckung der bei der Erection des männlichen Gliedes wirksamen Arterien bei dem Menschen und den Tieren." *Archiv für Anatomie, Physiologie und Wissenschaftliche Medicin* 1835: 202–13.

– 1837/40. *Handbuch der Physiologie des Menschen für Vorlesungen.* 2 Vols. Coblenz: Verlag von J. Hölscher.

– 1857. "Ueber die Fische, welche Töne von sich geben und die Entstehung dieser Töne." *Archiv für Anatomie, Physiologie und Wissenschaftliche Medicin* 1857: 249–79.

Müller-Schwarze, Dietland. 2006. *Chemical Ecology of Vertebrates.* Cambridge and New York: Cambridge University Press.

Mulloney, Brian, and Carmen Smarandache. 2000. "Fifty years of CPGs: two neuroethological papers that shaped the course of neuroscience." *Frontiers in Behavioral Neuroscience* 4: 1–8.

Munk, Ole. 1965. "The eyes of three benthic deep-sea fishes caught at great depths." *Galathea Report* 7: 137–49.

– 1966. "Ocular degeneration in deep-sea fishes." *Galathea Report* 8: 21–32.

– 1977. "The visual cells and retinal tapetum of the foveate deep-sea fish *Scopelosaurus lepidus* (Teleostei)." *Zoomorphologie* 87: 21–49.

Mykles, Donald L., Cameron K. Ghalambor, Jonathan H. Stillman, and Kars Tomanek. 2010. "Grand challenges in comparative physiology: integration across disciplines and across levels of biological organization." *Integrative and Comparative Biology* 50: 6–16.

Needham, Joseph. 1930. "The biochemical aspect of the recapitulation theory." *Biological Reviews* 5: 142–58.

Needham, Joseph, and Dorothy M. Needham. 1932. "Biochemical evidence regarding the origin of vertebrates." *Science Progress in the Twentieth Century* 26: 626–42.

Needham, Dorothy M., Joseph Needham, Ernest Baldwin, and John Yudkin. 1932. "A comparative study of the phosphagens, with special remarks on the origin of vertebrates." *Proceedings of the Royal Society of London B*, 10: 260–94.

Nicholls, John G., and David Van Essen. 1974. "The nervous system of the leech." *Scientific American* 230(1): 38–48.

Nolf, Pierre. 1937. "Notice sur Léon Fredericq." *Annuaire de l'Académie Royale de Belgique*, (1937): 47–100.

Novick, Alvin. 1959. "Acoustic orientation in the cave swiftlet." *Biological Bulletin* 117:497–53.

Nutton, Vivian. 2012. "*Physiologia* from Galen to Jacob Bording." Pages 27–40 in *Blood, Sweat and Tears: The Changing Concepts of Physiology from Antiquity into Early Modern Europe*, edited by Manfred Horstmanshoff, Helen King, and Claus Zittel. Leiden and Boston: Brill.

O'Malley, Charles D. 1964. *Andreas Vesalius of Brussels, 1514–1564*. Berkeley and Los Angeles: University of California Press.

Otis, Laura. 2007. *Müller's Lab*. New York, Oxford: Oxford University Press.

Pantin, Carl F.A. 1931. "The origin of the composition of the body fluids in animals." *Biological Reviews* 6: 459–82.

– 1932. "Physiological adaptation." *Zoological Journal of the Linnean Society* 37: 705–11.

Passino, Kevin M. 2005. *Biomimicry for Optimization, Control, and Automation*. London: Springer.

Paul, Harry W. 1985. *From Knowledge to Power: The Rise of the Science Empire in France, 1860–1939*. Cambridge, UK: Cambridge University Press.

Pauly, Philip J. 1987. *Controlling Life: Jacques Loeb and the Engineering Ideal in Biology*. New York: Oxford University Press.

Pearce, John M.S. 2009. "Marie-Jean-Pierre Flourens (1794–1867) and cortical localization." *European Neurology* 61: 311–14.

Pepperberg, Irene M. 1999. *The Alex Studies: Cognitive and Communicative Abilities of Grey Parrots*. Cambridge, Massachusetts: Harvard University Press.

Piccolino, Marco, and Marco Bresadola. 2013. *Shocking Frogs: Galvani, Volta, and the Electric Origins of Neuroscience*. Oxford and New York: Oxford University Press.

Pickford, Grace E., and James W. Atz. 1957. *The Physiology of the Pituitary Gland of Fishes*. New York: New York Zoological Society.

Pickstone, John V. 1981. "Bureaucracy, liberalism and the body in post-revolutionary France: Bichat's physiology and the Paris School of Medicine." *History of Science* 19: 115–42.

Pierce, George W., and Donald R. Griffin. 1938. "Experimental determination of supersonic notes emitted by bats." *Journal of Mammalogy* 19: 454–5.

Pietsch, Theodore W. 2005. "Dimorphism, parasitism, and sex revisited: modes of reproduction among deep-sea ceratioid anglerfishes (Teleostei: Lophiiformes)." *Ichthyological Research* 52: 207–36.

Pond, Caroline. 2009. "Dr Margaret (Peggy) Varley, pioneer ecophysiologist and ichthyologist." *Fish and Fisheries* 10: 359–60.

Postma, N., and P. Smit, editors. 1980. *Hermann Jacques Jordan (1877–1943. Nederlands eerste vergelijkend fysioloog*. Nijmegen: Katholieke Universiteit.

Poulson, Thomas L. 1963. "Cave adaptation in amblyopsid fishes." *American Midland Naturalist* 70: 257–90.

Prestrud, Pål. 1991. "Adaptations by the Arctic fox (*Alopex lagopus*) to the polar winter." *Arctic* 44: 132–8.

Price, David A., and Michael J. Greenberg. 1977. "Structure of a molluscan cardio-excitatory neuropeptide." *Science* 197: 670–1.

Priede, Imants G. 2017. *Deep-Sea Fishes: Biology, Diversity, Ecology and Fisheries*. Cambridge, UK: Cambridge University Press.

Prosser, C. Ladd. 1933a. "Correlation between development of behavior and neuromuscular differentiation in embryos of Eisenia foetida, Sav." *Journal of Comparative Neurology* 58: 603–41.

– 1933b. "Effect of the central nervous system on responses to light in Eisenia foetida Sav.." *Journal of Comparative Neurology* 59: 61–91.

– 1934a. "Action potentials in the nervous system of the crayfish. I. Spontaneous impulses" *Journal of Cellular and Comparative Physiology* 4: 185–209.

– 1934b. "Action potentials in the nervous system of the crayfish. II. Responses to illumination of the eye and caudal ganglion." *Journal of Cellular and Comparative Physiology* 4: 363–77.

– editor. 1950. *Comparative Animal Physiology*. Philadelphia: W.B. Saunders.

– 1969. "Principles and general concepts of adaptation." *Environmental Research* 2: 404–16.

– 1986. "The making of a comparative physiologist." *Annual Review of Physiology* 48: 1–6.

– editor. 1991. *Comparative Animal Physiology*, Parts A & B. Fourth Edition. New York, Wiley & Sons.

Przybos, Julia. 2009. "Balzac et l'imagination physiologique." *Lingua Romana* 8(1), Fall 2009.

Purpura, Dominick P. 1998. "Berta V. Scharrer 1906–1995." *Biographical Memoirs of the National Academy of Sciences* 74: 3–21.

Quatrefages, Armand de. 1854. *Souvenirs d'un naturaliste*. Paris: Charpentier, libraire-éditeur.

– 1862. *Physiologie comparée: métamorphoses de l'homme et des animaux*. Paris: J.B. Baillière et Fils.

– 1865. *Histoire naturelle des annelés marins et d'eau douce*. Paris: Librairie Encyclopédique de Roret.

– 1877. *L'espèce humaine*. Paris: Germer Baillière et Cie.

Randall, David. 2014. "Hughes and Shelton: the fathers of fish respiration." *Journal of Experimental Biology* 217: 3191–2.

Randall, David, Warren Burggren, and Kathleen French. 1997. *Eckert Animal Physiology: Mechanisms and Adaptations*. New York: W.H. Freeman and Company.

Rapport, Maurice M., Arda A. Green, and Irvine H. Page. 1948. "Crystalline serotonin." *Science* 108: 329–30.

Redfield, Alfred C. 1918. "The physiology of the melanophores of the horned toad *Phrynosoma*." *Journal of Experimental Zoology* 26: 275–333.

Redfield, Alfred C., and Marcel Florkin. 1931. "The respiratory function of the blood of *Urechis caupo*. *Biological Bulletin* 61: 185–210.

Regan, C. Tate. 1925. "Dwarfed males parasitic on the females in oceanic anglerfishes (Pediculati Ceratioidea)." *Proceedings of the Royal Society of London* B, 97: 385–400.

Renault, Marion. 2019. "Modern medicine still needs leeches." *Popular Science*, 8 July 2019.

Retzius, Gustaf. 1890. "Zur Kenntniss des Nervensystems der Crustaceen. I. Das centrale Nervensystem." *Biologische Untersuchungen*, Neue Folge 1: 1–50.

Revelle, Roger. 1995. "Alfred C. Redfield 1890–1983." *Biographical Memoirs of the National Academy of Sciences* 67: 315–29.

Rheinberger, Hans–Jörg. 2000. "Ephestia: the experimental design of Alfred Kühn's physiological developmental genetics." *Journal of the History of Biology* 33: 535–76.

Rice, Gillian. 1987. "The Bell-Magendie-Walker controversy." *Medical History* 31: 190–200.

Richards, Robert J. 1987. *Darwin and the Emergence of Evolutionary Theories of Mind and Behavior*. Chicago: University of Chicago Press.

Riddle, Oscar. 1935. "Contemplating the hormones." *Endocrinology* 19: 1–13.

Riddle, Oscar, Robert W. Bates, and Simon W. Dykshorn. 1933. "The preparation, identification and assay of prolactin – a hormone of the anterior pituitary." *American Journal of Physiology* 105: 191–216.

Riese, Walther, and George Arrington. 1963. "The history of Johannes Müller's doctrine of the specific energies of the senses: original and later versions." *Bulletin of the History of Medicine* 37: 179–83.

Robertson, Robert M., and Maurice Moulins. 1981. "Oscillatory command input to the motor pattern generators of the crustacean stomatogastric ganglion I. The pyloric rhythm." *Journal of Comparative Physiology* A, 143: 453–63.

Rogers, Charles G. 1927. *Textbook of Comparative Physiology*. New York: McGraw-Hill.

Rogers, Charles G., and Elsie M. Lewis. 1916. "The relation of the body temperature of certain cold-blooded animals to that of their environment." *Biological Bulletin* 31: 1–15.

Romanes, George J. 1882. *Animal Intelligence*. London: Kegan Paul, Trench & Co.

– 1883. *Mental Evolution in Animals*. London: Kegan Paul, Trench & Co.

Romero, Aldemaro. 2009. *Cave Biology: Life in Darkness*. Cambridge and New York: Cambridge University Press.

Ross, Donald M. 1981. "Illusion and reality in comparative physiology." *Canadian Journal of Zoology* 59: 2151–8.

Rostand, Jean. 1951. *Les origines de la biologie expérimentale et l'abbé Spallanzani*. Paris: Fasquelle.

Roth, Jesse, Derek LeRoith, et al. 1986. "The evolutionary origins of intercellular communication and the Maginot Lines of the mind." *Annals of the New York Academy of Sciences* 463: 1–11.

Rothschild, Lynn J., and Rocco L. Mancinelli. 2001. "Life in extreme environments." *Nature* 409: 1092–101.

Russell, Edward S. 1916. *Form and Function: A Contribution to the History of Animal Morphology*. London: J. Murray.

Russell, Frederick S. 1968. "Carl Frederick Abel Pantin 1899–1967." *Biographical Memoirs of Fellows of the Royal Society* 14: 417–34.

Saemundsson, Bjami. 1922. "Zoologiske Meddelelser fra Island. XIV. Il Fiske, ny for Island, og supplerendeOplysningerom andre, tidligere kendte." *Videnskabelige Meddelelser fra Dansk naturhistorik Forening i Kjøbenhavn* 74: 159–201.

Sarkowski, Heinz. 1996. *Springer-Verlag: History of a Scientific Publishing House*. Berlin, Heidelberg, New York: Springer-Verlag.

Schaller, Friedrich. 1985. "Wolfgang von Buddenbrock (1884–1964), der Zoologe und Physiologe." *Medizinhistorisches Journal* 20: 109–19.

Scharrer, Ernst. 1928. "Die Lichtempfindlichkeit blinder Elritzen." *Zeitschrift für vergleichende Physiologie* 7: 1–38.

Scharrer, Ernst, and Berta Scharrer. 1937. "Über Drüsen-Nervenzellen und neurosekretorische Organe bei Wirbellosen und Wirbeltieren." Biological Reviews 12: 185–216.

Scheer, Bradley T., editor. 1957. *Recent Advances in Invertebrate Physiology*. Eugene, Oregon: University of Oregon Pubications.

Schiller, Joseph. 1968. "Physiology's struggle for independence in the first half of the nineteenth century." *History of Science* 7: 64–87.

Schmidt-Nielsen, Bodil. 1995. *August and Marie Krogh: Lives in Science*. New York: Springer-Verlag.

Schmidt-Nielsen, Bodil, Knut Schmidt-Nielsen, Adelaide Brokaw, and Howard Schneiderman. 1948. "Water conversation in desert rodents." *Journal of Cellular and Comparative Physiology* 32: 331–60.

Schmidt-Nielsen, Bodil, Knut Schmidt-Nielsen, T.R. Houpt, and S.A. Jarnum. 1956. "Water balance of the camel." *American Journal of Physiology* 185: 185–94.

Schmidt-Nielsen, Knut. 1960. *Animal Physiology*. Englewood Cliffs, New Jersey: Prentice-Hall.

– 1972. *How Animals Work*. Cambridge, UK: Cambridge University Press.

– 1975. *Animal Physiology: Adaptation and Environment*. New York: Cambridge University Press.

– 1987. "Per Scholander 1905–1980." *Biographical Memoirs of the National Academy of Sciences* 57: 387–412.

– 1998. *The Camel's Nose: Memoirs of a Curious Scientist.* Washington, D.C., Covello, CA: Island Press/Shearwater Books.

Schmidt-Nielsen, Knut, C. Barker Jörgensen, and Humio Osaki. 1958. Extrarenal salt excretion in birds." *American Journal of Physiology* 193: 101–7.

Schmidt-Nielsen, Knut, Bodil Schmidt-Nielsen, S.A. Jarnum, and T.R. Houpt. 1956. "Body temperature of the camel and its relation to water economy." *American Journal of Physiology* 188: 103–12.

Schoffeniels, E. 1960. "Origine des acides aminés intervenant dans la régulation de la pression osmotique intracellulaire." *Archives Internationales de Physiologie et Biochimie* 68: 696–8.

Scholander, Per F. 1940. "Experimental investigations on the respiratory function in diving mammals and birds." *Hvalradets Skrifter* 22: 1–131.

– 1955. "Evolution of climatic adaptation in homeotherms." *Evolution* 9: 15–26.

– 1978. "Rhapsody in science." *Annual Review of Physiology* 40: 1–17.

Schultz, Stanley G. 2002. "William Harvey and the circulation of the blood: The birth of a scientific revolution and modern physiology." *News in Physiological Sciences* 17: 175–80.

Schwann, Theodor. 1839. *Mikroskopische Untersuchungen über die Übereinstimmung in der Struktur und dem Wachstum der Tiere und Pflanzen.* Berlin: Verlag der Sander'schen Buchhandlung (G.E. Reimer).

Scoville, William B. and Brenda Milner. 1957. "Loss of recent memory after bilateral hippocampal lesions." *Journal of Neurology, Neurosurgery and Psychiatry* 20: 11–21.

Selverston, Allen I., David F. Russell, and John P. Miller. 1976. "The stomatogastric nervous system: structure and function of a small neural network." *Progress in Neurobiology* 7: 215–90.

Selverston, Allen I., and John P. Miller. 1982. "Application of a cell inactivation technique to the study of a small neural network." *Trends in Neuroscience*, April 1982: 120–3.

Selye, Hans. 1936. "A syndrome produced by diverse nocuous agents." *Nature* 138: 32.

– 1938. "Adaptation energy." *Nature* 141: 1926.

Shackelford, Jole. 2003. *William Harvey and the Mechanics of the Heart.* Oxford, New York: Oxford University Press.

Sherman, R.G., and Harold L. Atwood. 1971. "Synaptic facilitation: long-term neuromuscular facilitation in crustaceans." *Science* 171: 1248–50.

Sherrington, Charles S. 1946. *The Endeavour of Jean Fernel: With a List of the Editions of His Writings.* Cambridge: Cambridge University Press.

Shiriagin, Vladimir, and Sigrun I. Korsching. 2019. "Massive expansion of bitter taste receptors in blind cavefish, *Astyanax mexicanus*." *Chemical Senses* 44: 23–32.

Shoja, Mohammadali M., R. Shane Tubbs, Marios Loukas, Ghaffar Shokouhi, and Mohammad R. Ardalan. 2008. "Marie-François Xavier Bichat (1771–1802) and his contributions to the foundations of pathological anatomy and modern medicine." *Annals of Anatomy* 190: 413–20.

Sillard, Jean-Christophe. 1979. "Quatrefages et le transformisme." *Revue de synthèse* 95/96: 283–95.

Singer, Charles. 1957. *A Short History of Anatomy & Physiology From the Greeks to Harvey.* New York: Dover Publications.

Singer, Michael A. 2007. *Comparative Physiology, Natural Animal Models and Clinical Medicine: Insights into Clinical Medicine from Animal Adaptations.* London: Imperial College Press.

Slack, Nancy G. 2003. "Are research schools necessary? Contrasting models of 20th century research at Yale led by Ross Granville Harrison, Grace E. Pickford and G. Evelyn Hutchinson." *Journal of the History of Biology* 36: 501–29.

Smith, Carol. 2003. "Aubrey Gorbman: renowned uw zoologist 'lived, breathed science.'" *Seattle Post-Intelligencer*, 29 September 2003.

Smith, Kenneth L., Jr, and R.J. Baldwin. 1997. "Laboratory and in situ methods for studying deep-sea fishes." Pages 351–77 in *Fish Physiology, Vol. 16: Deep-Sea Fishes.* Edited by David J. Randall and Anthony P. Farrell. San Diego, California: Academic Press.

Somero, George N. 1969. "Pyruvate kinase variants of the Alaskan king-crab. Evidence for a temperature-dependent interconversion between two forms having distinct and adaptive kinetic properties." *Biochemical Journal* 114: 237–41.

– 2009. "Clifford Ladd Prosser, May 12, 1907–February 3, 2002." *Biographical Memoirs of the National Academy of Science* 91: 243–57.

Somero, George N., and Arthur L. De Vries. 1967. "Temperature tolerance of some Antarctic fishes." *Science* 156: 257–8.

Somero, George N., and Raul C. Suarez. 2005. "Peter Hochachka: adventures in biochemical adaptation." *Annual Review of Physiology* 67: 25–37.

Somero, George N., Arthur C. Giese, and Donald E. Wohlschlag. 1968. "Cold adaptation of the Antarctic fish *Trematomus bernacchii*." *Comparative Biochemistry and Physiology* 26: 223–33.

Sonntag, Otto. 1974. "The motivations of the scientist: the self–image of Albrecht von Haller." *Isis* 65: 336–51.

Spallanzani, Lazaro. 1794. *Lettere sopra il Sospetto di un Nuovo Senso nei Pipistrelli ... Con le Risposte dell'Abate Antonmaria Vassalli*. Turin: Stamperia Reale.

Speidel, Carl C. 1922. "Further comparative studies in other fishes of cells that are homologous to the large irregular glandular cells in the spinal cord of the skates." *Journal of Comparative Neurology* 34: 303–17.

Spencer, Herbert. 1852. "The development hypothesis. Haythorne Papers no. 2." *The Leader* 3, 104: 280–1.

Stanzione, Massimo. 2011. "Developing science at the risk of oblivion: the case of Filippo Bottazzi." *European Review* 19: 445–67.

Steiger, Sandra, Thomas Schmitt, and H. Martin Schaefer. 2011. "The origin and dynamic evolution of chemical information transfer." *Proceedings of the Royal Society of London* B, 278: 970–9.

Stendhal. 1927. *Souvenirs d'égotisme*. Paris: Le Divan.

Storey, Kenneth B. 1974. "Metabolic Consequences of Diving: The Anoxic Turtle." PhD Dissertation, University of British Columbia.

Storey, Kenneth B., and Janet M. Storey. 1988. "Freeze tolerance in animals." *Physiological Reviews* 68: 27–84.

Sturtevant, Alfred H. 1959. "Thomas Hunt Morgan 1866–1945." *Biographical Memoirs of the National Academy of Sciences* 33: 283–325.

Suarez, Raul K., and David R. Jones. 2002. "Peter W. Hochachka (1937–2002)." *Nature* 420: 140.

Takeuchi, Ayako, and Noriko Takeuchi. 1964. "The effect on crayfish muscle of iontophoretically applied glutamate." *Journal of Physiology* 170: 296–317.

Tarttelin, Emma E., Elena Frigato, et al. 2012. "Encephalic photoreception and phototactic response in the troglobiont Somalian blind cavefish *Phreatichthys andruzzii*." *Journal of Experimental Biology* 215: 2898–903.

Tauc, Ladislav. 1957. "Les divers modes d'activité du soma neuronique ganglionnaire de l'Aplysie et de l'Escargot." Pages 91–119 in *Microphysiologie des éléments excitables*, edited by A. Fessard. Paris: CNRS Editions.

Tchernavin, Vladimir V. 1935. *I Speak for the Silent: Prisoners of the Soviets*. Boston and New York: Hale, Cushman & Flint.

– 1953. *The Feeding Mechanisms of a Deep-Sea Fish, Chauliodus sloani Schneider*. London: British Museum of Natural History.

Ternaux, Jean-Pierre, and François Clarac. 2012. *Le Bestiaire Cérébral*. Paris: CNRS Editions.

Thinès, Georges. 1953. "Recherches expérimentales sur la photosensibilité du poisson aveugle *Caecobarbus geertsii* Blgr." *Annales de la Société Royale Zoologique de Belgique* 84: 231–65.

– 1955. "Les poissons aveugles. I. Origine, taxonomie, répartition géographique, comportement." *Annales de la Société Royale Zoologique de Belgique* 86: 5–128.

Thorpe, William H. 1983. "Karl von Frisch. 20 November 1886–12 June 1982. *Biographical Memoirs of Fellows of the Royal Society* 29: 196–200.

Tipton, Jason A. 2014. *Philosophical Biology in Aristotle's Parts of Animals*. Heidelberg, New York, Dordrecht, London: Springer, Cham.

Treviranus, Gottfried R. 1802. *Biologie: oder Philosophie der lebenden Natur für Naturforscher und Aerzte*. Göttingen: J.F. Röwer.

Trewavas, Ethelwynn. 1949. "Dr. Vladimir Tchernavin." *Nature* 163: 755–6.

Tubbs, R. Shane. 2015. "Anatomy is to physiology as geography is to history; it describes the theatre of events." *Clinical Anatomy* 28: 151.

Tubbs, R. Shane, Marios Loukas, Mohammadali M. Shoja, Mohammad R. Ardalan, and W. Jerry Oakes. 2008. "Richard Lower (1631–1691) and his early contributions to cardiology." *International Journal of Cardiology* 128: 17–21.

Twarog, Betty M. 1954. "Responses of a molluscan smooth muscle to acetylcholine and 5–hydroxytryptamine." *Journal of Cellular and Comparative Physiology* 44: 141–63.

Twarog, Betty M., and Irvine H. Page. 1953. "Serotonin content of some mammalian tissues and urine and a method for its determination." *American Journal of Physiology* 175: 157–61.

Usherwood, Peter N.R. 1977. "Neuromuscular transmission in insects." Pages 31–48 in *Identified Neurons and Behavior of Arthropods*, edited by Graham Hoyle. New York: Plenum Press.

Valenstein, Elliot S. 2005. *The War of the Soups and the Sparks*. New York: Columbia University Press.

Vandel, Albert. 1965. *Biospeleology: The Biology of Cavernicolous Animals*. Oxford, UK: Pergamon Press.

Vartanian, Ariam. 1960. *La Mettrie's l'Homme Machine. A Study in the Origins of an Idea*. Princeton, New Jersey: Princeton University Press.

Verworn, Max. 1899. *An Outline of the Science of Life*. London, New York: Macmillan & Co.

Vialli, Maffo, and Vittorio Erspamer. 1937. "Ricerche sul secreto delle cellule enterocromaffini. IX. Intorno alla natura chimica della sostanza specifica." *Bolletano della Societa medica–chirurgica di Pavia* 51: 1111–16.

Virey, Julien J. 1825. *De la puissance vitale considérée dans ses fonctions physiologiques chez l'homme et tous les êtres organisés*. Paris: Crochard Libraire.

Vogel, Steven. 2008. "Knut Schmidt-Nielsen, September 24, 1915–January 25, 2007." *Biographical Memoirs of the Royal Society* 54: 319–31.

Volta, Alessandro. 1800. "On the electricity excited by the mere contact of conducting substances of different kinds." *Transactions of the Royal Society of London* 90: 403–31.

– 1816. *Collezione dell'opere del cavaliere Conte Alessandro Volta*. Edited by Patrizio Comasco. Volume 2, Part 1. Firenze: Nella stamperia di G. Piatti.

Vulpian, Edmé F.A. 1888. "Eloge historique de M. Flourens." *Mémoires de l'Académie des sciences de l'Institut de France*, deuxième série, 44: cl–clxxxiv.

Vyleta, Daniel M. 2007. *Crime, Jews and News. Vienna 1895–1914*. New York/ Oxford: Berghahn Books.

Walker, Robert J. 2004. "Professor Gerald Allan Kerkut: 19 August 1927–6 March 2004: Editor: *Comparative Biochemistry and Physiology* 1960–1994." *Comparative Biochemistry and Physiology* A, 138: 261–2.

Walker, Theodore J., and Arthur D. Hasler. 1949. "Detection and discrimination of odors of aquatic plants by the bluntnose minnow (*Hyborhynchus notatus*)." *Physiological Zoology* 22: 45–63.

Watanabe, Akira, and Theodor H. Bullock. 1960. "Modulation of activity of one neuron by subthreshold slow potentials in another in lobster cardiac ganglion." *Journal of General Physiology* 43: 1031–45.

Waterman, Talbot H. 1975. "Expectation and achievement in comparative physiology." *Journal of Experimental Zoology* 194: 309–44.

Watkins, Jeffrey C., and David E. Jane. 2006. "The glutamate story." *British Journal of Pharmacology* 147: S100–8.

Weber, Hans H., and Hans H. Loeschcke. 1964. "Hans Winterstein." *Ergebnisse der Physiologie* 55: 1–26.

Wehner, Rüdiger, A.C. Marsh, and S. Wehner. 1992. "Desert ants on a thermal tightrope." *Nature* 357: 586–7.

Weiger, Wendy A. 1997. "Serotonergic modulation of behaviour: a phylogenetic overview." *Biological Reviews* 72: 61–95.

Welch, G. Rickey. 2008. "In retrospect: Fernel's *Physiologia*." *Nature* 456: 446–7.

Wells, George P. 1976. "The early days of the S.E.B." Pages 1–6 in *Perspectives in Experimental Biology*. London, New York: Pergamon Press.

– 1978. "Lancelot Thomas Hogben, 9 December 1895 – 22 August 1975." *Biographical Memoirs of Fellows of the Royal Society* 24: 183–221.

West, John B. 2013a. "Marcello Malpighi and the discovery of the pulmonary capillaries and alveoli." *American Journal of Physiology* 304: L383–90.

– 2013b. "Torricelli and the ocean of air: the first measurement of barometric pressure." *Physiology* 28: 66–73.

– 2015. *Essays on the History of Respiratory Physiology*. New York, Heidelberg, Dordrecht, and London: Springer.

Whitaker-Azmitia, Patricia M. 1999. "The discovery of serotonin and its role in neuroscience." *Neuropsychopharmacology* 21: 2S–8S.

Wienecke, Barbara 2010. "The history of the discovery of emperor penguin colonies, 1902–2004." *Polar Research* 46: 271–6.

Wiersma, Cornelis A.G. 1933. "Vergleichende Untersuchungen über das periphere Nerven-Muskelsystem von Crustaceen." *Zeitschrift für vergleichende Physiologie* 19: 349–85.

– 1947. "Giant nerve fibre system of the crayfish: A contribution to comparative physiology of synapse." *Journal of Neurophysiology* 10: 23–38.

Wiersma, Cornelis A.G., and Graham M. Hughes. 1961. "On the functional anatomy of neuronal units in the abdominal cord of the crayfish, *Procambarus clarkii* (girard)." *Journal of Comparative Neurology* 116: 209–28.

Wiersma, Cornelis A.G., and K. Ikeda. 1964. "Interneurons commanding swimmeret movements in the crayfish, *Procambarus clarki* (girard)." *Comparative Biochemistry and Physiology* 12: 509–16.

Willmer, Pat, Graham Stone, and Ian Johnston. 2005. *Environmental Physiology of Animals*, 2nd edition. Oxford: Blackwell Publishing.

Willows, A.O. Dennis, D.A. Dorsett, and Graham Hoyle. 1973. "The neuronal basis of behavior in *Tritonia*. I. Functional organization of the central nervous system." *Journal of Neurobiology* 4: 207–37.

Wilson, Donald M. 1961. "The central nervous control of flight in a locust." *Journal of Experimental Biology* 38: 471–90.

Winchester, Simon. 2009. *The Man Who Loved China: The Fantastic Story of the*

Eccentric Scientist Who Unlocked the Mysteries of the Middle Kingdom. New York: HarperCollins.

Windle, William F. 1975. "Clarence Luther Herrick and the beginning of neuroscience in America." *Experimental Neurology* 49(1/2): 1–10.

Winterstein, Hans. 1911. "Die automatische Tätigkeit der Atemzentren." *Pflügers Archiv für gesammte Physiologie* 138: 157–66.

– editor. 1910/25. *Handbuch der vergleichenden Physiologie*. Jena: Gustav Fischer.

Wise, Peter H. 2011. *A Matter of Doubt: The Novel of Claude Bernard*. CreateSpace Independent Publishing Platform.

Wittlinger, Matthias, Rüdiger Wehner, and Hrald Wolf. 2007. "The desert ant odometer: a stride integrator that accounts for stride length and walking speed." *Journal of Experimental Biology* 210: 198–207.

Wojtowicz, J. Martin, B.R. Smith, and Harold L. Atwood. 1991. "Activity-dependent recruitment of silent synapses." *Annals of the New York Academy of Sciences* 627: 169–79.

Woodman, Alice S. 1939. "The pituitary gland of the Atlantic salmon." *Journal of Morphology* 65: 411–35.

Wright, James R., Jr. 2002. "From ugly fish to conquer death: J J R Macleod's fish insulin research, 1922–24." *The Lancet* 359: 1238–42.

Wright, Thomas. 2012. *Circulation: William Harvey's Revolutionary Idea*. London: Chatto & Windus.

– 2013. "The art of medicine: William Harvey goes back to the future." *The Lancet* 381: 620–1.

Wyatt, Tristram D. 2009. "Pheromones and other chemical communication in animals." *Encyclopedia of Neuroscience*, 2009, 611–16.

Wynne-Edwards, Vero C. 1962. *Animal Dispersion in Relation to Social Behaviour*. Edinburgh: Oliver and Boyd.

Yasumoto, T., and M. Murata. 1993. "Marine toxins." *Chemistry Review* 93: 1897–1909.

Yoshizawa, Masato, William R. Jeffery, Sietse M. van Netten, and Matthew J. McHenry. 2014. "The sensitivity of lateral line receptors and their role in the behavior of Mexican blind cavefish (*Astyanax mexicanus*)." *Journal of Experimental Biology* 217: 886–95.

Yonge, C.M. 1923. "Studies on the comparative physiology of digestion. I. The mechanism of feeding, digestion, and assimilation in the lamellibranch *Mya*." *Journal of Experimental Biology* 1: 15–63.

Young, John Z. 1938. "The functioning of the giant nerve fibres of the squid."
   Journal of Experimental Biology 15: 170–85.

– 1966. *The Memory System of the Brain*. Berkeley and Los Angeles: University
   of California Press.

– 1985. "Cephalopods and neuroscience." *Biological Bulletin* 168: 153–8.

Zimmer, Heinz-Gerd. 1996. "Carl Ludwig: the man, his time, his influence."
   European Journal of Physiology 432: R9–R22.

Index

ॐ